Exploring Our Living Planet

National Geographic Society

Explor
Our Li
Planet

ing
ving

Exploring Our Living Planet

by Robert D. Ballard

Published by
The National Geographic Society

Gilbert M. Grosvenor
President

Melvin M. Payne
Chairman of the Board

Owen R. Anderson
Executive Vice President

Robert L. Breeden
Vice President, Publications and Educational Media

Prepared by
National Geographic Book Service

Charles O. Hyman
Director

Kenneth C. Danforth
Managing Editor

Anne Dirkes Kobor
Illustrations Editor

Staff for this book

Jonathan B. Tourtellot
Editor

Susan Eckert Sidman
Research Editor

David M. Seager
Art Director

Linda B. Meyerriecks
Picture Editor

Robert Arndt
Ross S. Bennett
Edward Lanouette
David F. Robinson
Margaret Sedeen
Editor-Writers

Jean Kaplan Teichroew
Art Coordination

Orren J. Alperstein
Carolyn H. Anderson
Paul A. Dunn
Suzanne P. Kane
Melanie Patt-Corner
Jean Kaplan Teichroew
Editorial Research

Charlotte Golin
Design Assistant

Greta Arnold
Illustrations Research

Jane S. Offen
Illustrations Assistant

Georgina L. McCormack
Teresita C. Sison
Editorial Assistants

Robert C. Firestone
Production Manager

Karen F. Edwards
Richard S. Wain
Asst. Production Managers

John T. Dunn
Ronald E. Williamson
Engraving and Printing

Photographs by
David Austen
Nathan Benn
Emory Kristof
George F. Mobley
James A. Sugar
and others

Paintings by
Robert Hynes
Davis Meltzer
Rob Wood
and others

Relief maps by
John A. Bonner
Timothy J. Carter
Arthur James Cox
John G. Leocha
Tibor G. Toth
Cartographic Division

Diagrams by
John D. Garst, Jr.
Judith Bell Siegel
Gary M. Johnson
Mark Seidler
Publications Art

Contributions by
Thomas B. Allen
Stan Benjamin
Mary B. Dickinson
Bruce A. Lewenstein
Diana E. McFadden
Shirley Scott
Lise M. Swinson

Jeffrey A. Brown
Index

First edition 240,000 copies
176 illustrations
97 paintings, diagrams, and maps

Contents

Foreword

We let the cliché roll off our tongues, "this fascinating world," when we see something that is new or surprising. But we seldom think about the magnificence of the world as a—well, as a *world*.

Only on those rare occasions of great volcanic eruptions or rending earthquakes do we remember that this Earth on which we walk and ride and dig and build is in a sense as much a living thing as the creatures it has spawned.

Earth is living and dying, all the time creating and destroying its own surface. It is a body in motion, alive—and in its vitality, awesome.

Compared to the great energy that Earth unleashes, humankind is insignificant. We haven't built a nuclear weapon with the explosive power of a Mount St. Helens; with all our chemicals and machines we are helpless midgets before the moving Earth.

But when I report on science for television, I can see that we are at least making great progress in understanding how our planet works. It has been but the flick of a geologic eyelid since we developed the theory of plate tectonics, the discovery that our Earth's surface is a series of moving plates, constantly grinding against each other or pulling apart from one another. Yet in that brief time we have found explanation for much of what was mystery in the past.

We know now why mountains have risen, why lakes and oceans and islands have appeared and disappeared. Earth's restlessness may help explain the plagues of Egypt, the legend of Atlantis, and the fates of the Minoan and Maya civilizations.

The scientists who are unlocking the secrets of our planet are contributing an exciting chapter to our history. Their findings themselves make fascinating reading, but the exploits by which they gather their data are also the stuff of which tales are spun.

Earth does not yield up its secrets easily. Expeditions in search of answers have dropped men into the steaming craters of volcanoes and the crushing depths of the oceans.

As much as our astronauts who walked the moon, the scientists who have walked the fiery lava fields of an erupting volcano have gone where no man preceded them, into an environment so hostile that life could be sustained only by the equipment with which they surrounded themselves.

And so with those who have ridden the tiny submersible *Alvin* to the smoky vents at the bottom of the Pacific Ocean. *Alvin* is to the deep ocean what the Apollo spacecraft was to deep space. They are similar in size and complexity —and lack of comfort, as I found out during my own *Alvin* dive. But discomfort is forgotten in the excitement of discovery, and in the realization that through these discoveries we will learn how to predict Earth's more destructive convulsions, and how better to harness the great energy inside this fascinating world.

WALTER CRONKITE

Spaceman on his own planet, an asbestos-clad volcanologist probes steaming ash in the crater of Indonesia's Merapi volcano.

answers from this silent rock,
the canyon slices time, not
space: Brink to base, it spans
1.7 billion years—still only
a fraction of Earth's life. But
we did not always know this. . . .

The poor world is almost six thousand years old.
Shakespeare AS YOU LIKE IT

On a human scale, the planet Earth seems to be a spinning rock frozen in time, unchanging and unchanged, where a day lasts hours and a lifetime rarely exceeds a century. But on a larger scale, where galaxies spin once in 200 million years, where a planet survives for billions, Earth is lively and dynamic. I am a geologist and to me our world is a living thing.

Until modern times we humans shared the perspective of a tiny mayfly fluttering about a giant sequoia. The mayfly, with an adult life measured in hours, does not perceive the 2,000-year-old tree to be alive. After all, the insect has lived there all its life and the tree has never moved.

We, scurrying around on an Earth that has been in existence billions of years, tend to think the same thing, that our rocky globe has no life of its own. But it has, and to understand Earth's crustal activity we must understand time—long, long expanses of time.

Ever since we humans began to count the days, we have attempted to measure the span of our existence. Early civilizations developed calendars to monitor the seasons. Ancient stone observatories, like the Maya temples in Mexico and the Indian medicine wheel in Wyoming, traced and predicted the movement of the heavens. But the big question always remained—when did Earth begin?

In the 17th century, Anglican Archbishop James Ussher developed a biblical chronology that put the date of creation at 9 a.m., October 26, in the year 4004 B.C. Though he bears the brunt of modern ridicule, Ussher was a serious and learned scholar, making the most of what he considered the best information available. His calculations were taken seriously for 200 years.

Scholars of Ussher's day assumed that Earth had formed from a molten ball that cooled, leaving a lumpy crust of mountains, valleys, plains, riverbeds, and oceans.

They also clung to the doctrine of Catastrophism, which held that most Earth changes occurred violently and rapidly—witness, they argued, the Great Flood and the volcanic eruption that buried Pompeii. Catastrophism also agreed comfortably with Ussher's 6,000-year time scale—catastrophes, sudden by definition, do not last long.

But evidence that Earth had a much older origin began to accumulate as scholars turned their attention from life *on* the planet to rocks *in* the planet.

Near the end of the 18th century many scientists—known as Neptunists—believed that a great globe-engulfing sea had laid down all the rocky layers of Earth. As the waters subsided, dry land emerged, exposing the layers. The youngest and topmost were the alluvial deposits: sand and gravel washed down from the mountains into valleys and lowlands. Abraham G. Werner, the German founder of Neptunism, ascribed the igneous intrusions found in sedimentary layers, as well as the eruption of volcanoes, to coal seams bursting into combustion. Though wrong about that, Werner was right about the age of Earth—he thought it was much older than humans.

A mere 6,000-year history for Earth also troubled the Scotsman James Hutton, the father of modern geology. Trained as a physician, Hutton chose instead to become a gentleman farmer, which brought him close to the land. He wondered about the erosion of soil and puzzled over rock formations, and eventually devoted his life to studying the history of our planet. In 1795 his *Theory of the Earth* proposed that Earth was in continuous but gradual change, constantly decaying, renewing, and repairing itself. He believed that Earth had "no vestige of a beginning, no prospect of an end." Hutton had sought evidence for his theories and found it in two important observations.

The wall and the stream

Across Northumberland ran Hadrian's Wall, a stone barrier begun by the Romans around A.D. 122. Hutton saw that the stones had suffered little deterioration from weathering. If 16 centuries could barely scar a 6-foot-high stone wall, then the time needed for erosion to wear down mountains and carry sand and gravel to the sea must be enormous.

Now he needed proof of deep, extended time, of worlds stacked upon worlds. Searching the countryside in 1787, he came upon a section of riverbank cut deep by the River Jed. There, revealed in cross section above the surface of the water, lay three worlds one upon another. Lowest in the bank were vertical beds of schist, a hard metamorphic rock. Their upper ends were worn away, beveled by open-air erosion. Above them lay

Layers of rock at right angles in a Scottish riverbank changed forever our view of Earth history. In 1787, when most scholars believed our planet to be only a few thousand years old, James Hutton used this formation on the River Jed to prove the awesome reach of geologic time. Both sets of rock layers had formed horizontally beneath an ocean. Yet in the years between, earth movement upended the bottom group into mountains. Erosion wore them down before the sea topped them off with the upper layers. Now all stood above water again, a calendar of stone millions of years long.

This long-lost contemporary drawing came to light in 1968.

horizontal layers of sandstone, and above the sandstone rested the young soil of the present.

Hutton reasoned that two sets of layered rock at an angle to each other, a formation geologists now call an angular unconformity, must have required hundreds of thousands of years to take shape—enough time for the schist to form, to tilt, and to erode before the sea deposited the sandstone on top of it. No single event or flood could have constructed what he saw. The formation defied both Catastrophist and Neptunist concepts of time, not to mention Ussher's 6,000 years. Little did Hutton realize how long it took just to bevel the layers of tilted schist—70 million years.

While Hutton sought other unconformities, William Smith, an English geologist, began arranging groups of fossils. He determined which fossils were peculiar to each parent rock layer. By this method he could estimate the sequence of rock layers or strata. His work, too, suggested slow, gradual change.

Relying on the work of Hutton and Smith, England's Charles Lyell put the Catastrophists' argument to final rest by publishing in 1830 his *Principles of Geology,* in which he supported Hutton's ideas of renewal by stating that the Earth of today behaves as it always has, a principle he named Uniformitarianism.

The attempt to establish Earth's exact birthday continues. One reason for our trips to the moon was to find rocks older than any remaining on our own planet's turbulent surface. At present, information from the moon, from Earth's most ancient rocks, and from meteorites, puts Earth's age at 4.6 billion years.

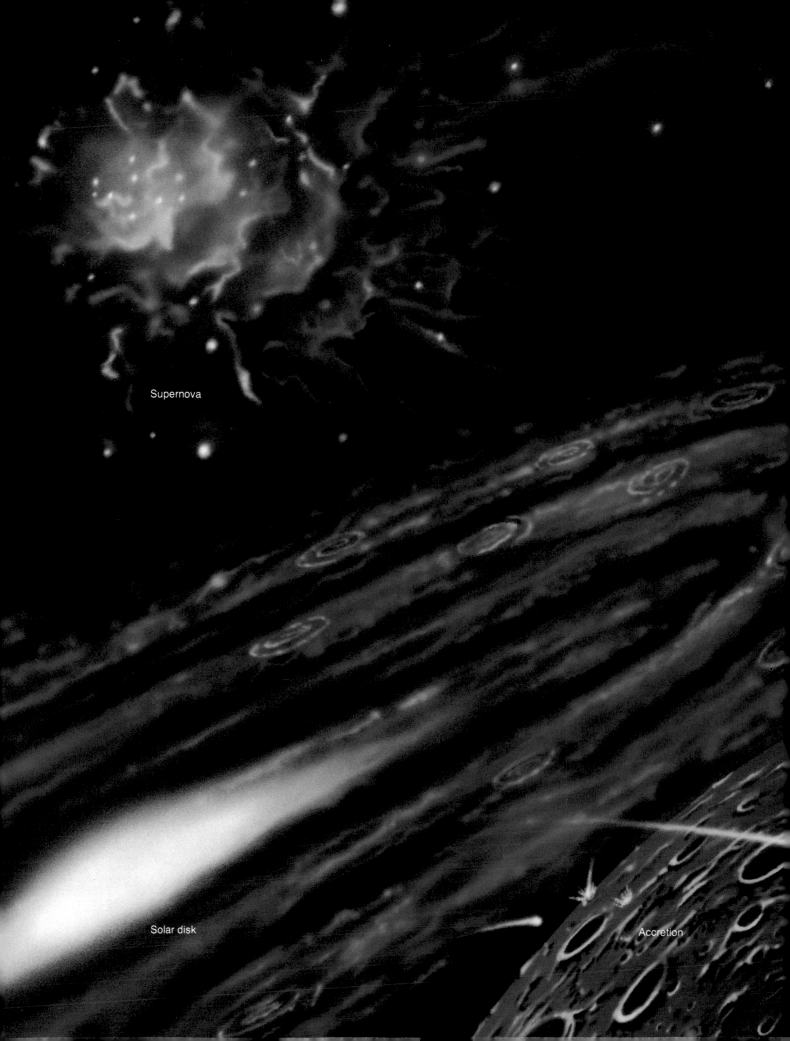

Supernova

Solar disk

Accretion

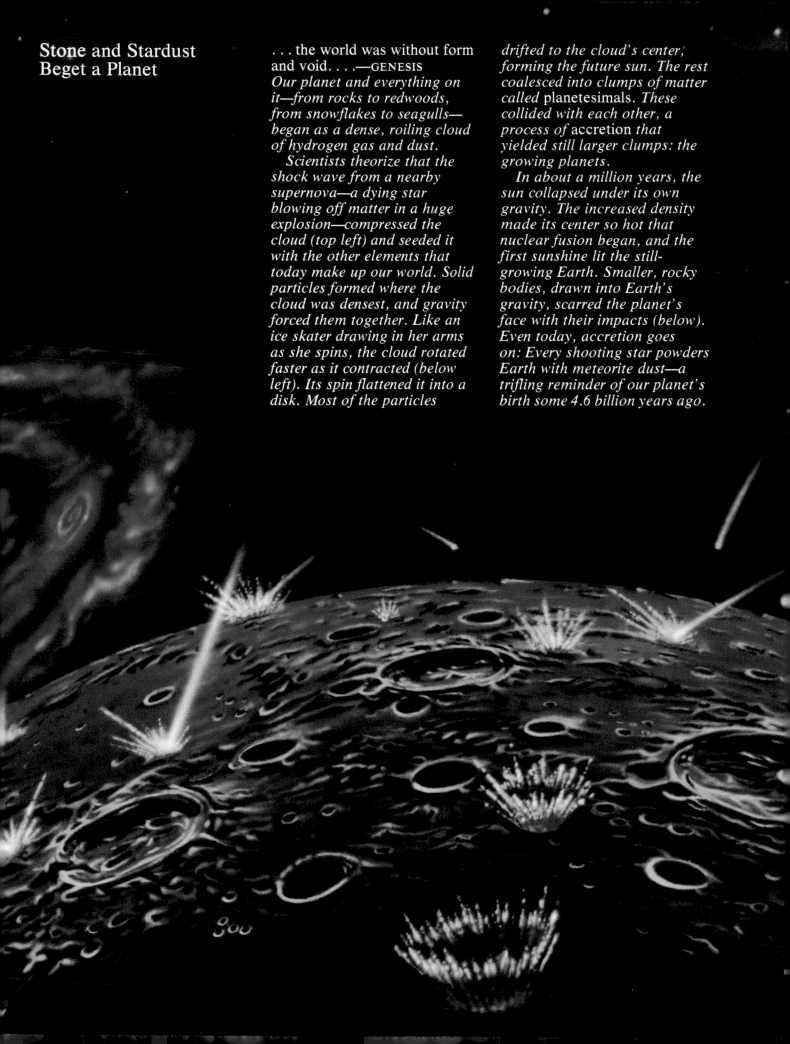

Stone and Stardust Beget a Planet

. . . the world was without form and void. . . .—GENESIS

Our planet and everything on it—from rocks to redwoods, from snowflakes to seagulls— began as a dense, roiling cloud of hydrogen gas and dust.

Scientists theorize that the shock wave from a nearby supernova—a dying star blowing off matter in a huge explosion—compressed the cloud (top left) and seeded it with the other elements that today make up our world. Solid particles formed where the cloud was densest, and gravity forced them together. Like an ice skater drawing in her arms as she spins, the cloud rotated faster as it contracted (below left). Its spin flattened it into a disk. Most of the particles drifted to the cloud's center, forming the future sun. The rest coalesced into clumps of matter called planetesimals. *These collided with each other, a process of* accretion *that yielded still larger clumps: the growing planets.*

In about a million years, the sun collapsed under its own gravity. The increased density made its center so hot that nuclear fusion began, and the first sunshine lit the still- growing Earth. Smaller, rocky bodies, drawn into Earth's gravity, scarred the planet's face with their impacts (below). Even today, accretion goes on: Every shooting star powders Earth with meteorite dust—a trifling reminder of our planet's birth some 4.6 billion years ago.

The Iron Catastrophe:
A Worldwide Meltdown

Almost a billion years after accretion began, a slow-motion cataclysm may have rebuilt the Earth from core to surface. Our planet today has a layered interior, but scientists have long disagreed about how it got that way. Some argue that the Earth was molten from the beginning, sorting itself into layers as it formed. But an opposing school holds that Earth began as a cold mass of rubble and later transformed itself into a planet that could someday harbor life. If so, we owe our existence to a

spectacular event called the Iron Catastrophe.

Within the rock of the young Earth, radioactive elements released heat as they decayed. Though weakened by time, they are still decaying today, even in ordinary granite. If you sealed a gallon-sized cube of granite in a perfectly insulated container with a pot of coffee—and left it for about 850,000 years—the granite would brew the coffee.

The heat of radioactive decay helped raise Earth's internal temperature to nearly 1,000°C. Impacts and gravity's pressure

4.5 billion years ago

1

2

3

added more heat as incoming planetesimals continued to pile new material on top of old, squeezing and warming the planet's interior (1). Eventually, pressure and temperature at a level 400 to 800 kilometers below Earth's surface reached the point where the iron in the rock there melted. The Iron Catastrophe was under way.

As rock and iron melted (2), drops and streams of the metal began to fall through the lighter rock, displacing it upward. The energy of this incredible ironfall heated the planet further. As temperatures mounted, the zone of melting broadened, spreading both downward and upward (3). As it neared Earth's crust volcanism began venting lava and gases on the surface. At last most of the original crust, floating unstably on liquid rock, melted and foundered (4). Earth was organizing itself into layers.

Heavy elements like nickel sank toward Earth's center with the iron and solidified under the enormous pressure there, forming the planet's inner core. Silicates and other lighter materials rose to the surface. Convection currents in the liquid rock carried heat to the surface as well, cooling the planet much the way stirring cools a hot bowl of soup. A new crust began to form. (The few fragments of this crust still visible today are the oldest terrestrial rocks we know, many in North America and Africa.)

Water vapor and other gases, once chemically bound inside the planet, now erupted from the surface in huge quantities (5), creating Earth's primitive oceans and atmosphere.

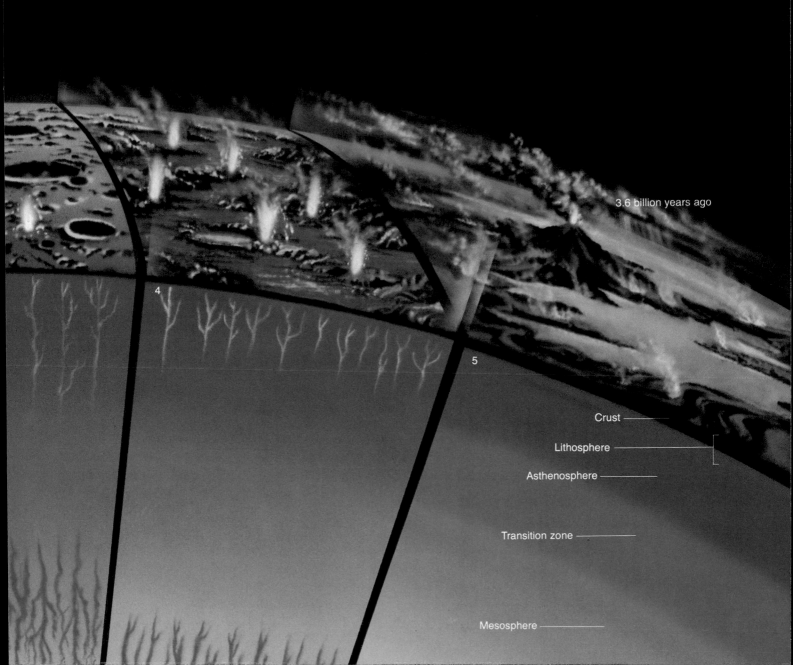

3.6 billion years ago

Crust

Lithosphere

Asthenosphere

Transition zone

Mesosphere

Life Remodels
The Globe

Earth's primitive atmosphere held no free oxygen. It was mostly water vapor, nitrogen, and carbon dioxide. The sun's deadly ultraviolet radiation blasted the planet. But water vapor meant rain, and rain washed sediment and chemicals into the sea, enriching it. The oceans were ready for life.

It soon appeared. Scientists have found evidence of simple cells in 3.5-billion-year-old rocks. But how could life begin on an irradiated, oxygen-starved Earth? Experiments reveal the answer.

In their labs, scientists duplicated Earth's primitive atmosphere. To simulate the lightning and harsh sunlight of early Earth, they seared their gas mixtures with high-voltage sparks and ultraviolet light— forces that make and break chemical bonds. In the resulting brew, they found amino acids, the building blocks of life. They believe the same process occurred on Earth 3 to 4 billion years ago.

Another experiment showed that amino acids on a hot, dry surface—like that of a cooling

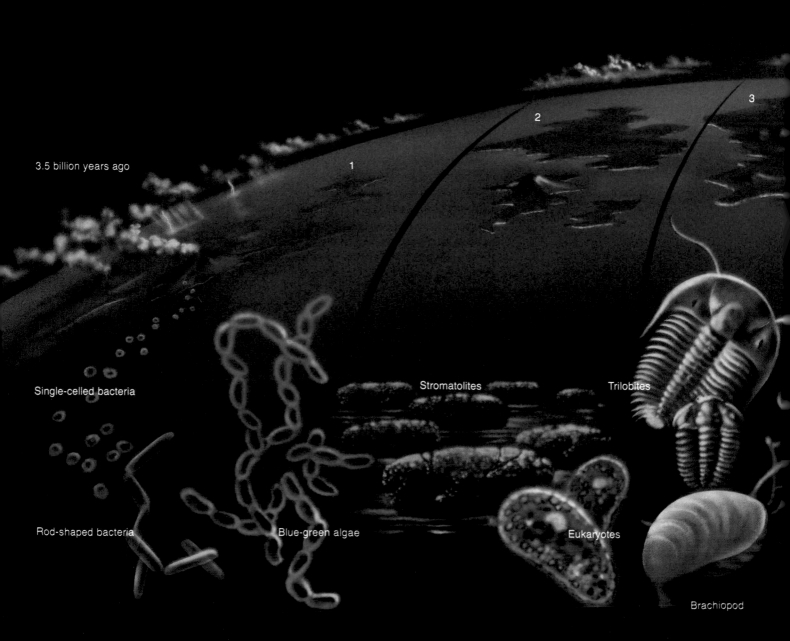

3.5 billion years ago

1

2

3

Single-celled bacteria

Stromatolites

Trilobites

Rod-shaped bacteria

Blue-green algae

Eukaryotes

Brachiopod

rock—form cell-like spheres when splashed with water. If rain washed the spheres into a tidal pool, a place safe from ultraviolet radiation, more complex molecules could form. At last one appeared with the ability to reproduce itself—a molecule similar to DNA, which exists today in every living cell. With its appearance, life had begun on Earth (1).

The early cells lived in the oceans and "fed" by absorbing organic matter. Then some found a better way: They began to photosynthesize, using the energy of sunlight to make their food. These first plants, the blue-green algae, formed colonies called stromatolites (2). More important, they also produced oxygen, a by-product of photosynthesis. With oxygen, living things began to change the planet that had borne them.

As green plants spread, they leaked free oxygen into the seas and atmosphere. Iron remaining in the crust rusted to iron oxide—a record in rock of oxygen's first appearance. New ocean life forms developed a way to use oxygen for turning food into energy, and that process—respiration—proved so efficient that oxygen users spread and diversified into many different species (3).

As atmospheric oxygen increased, oxygen molecules linked to form ozone in the stratosphere. The ozone layer screened out much of the sun's lethal ultraviolet light, and for the first time dry land was safe for life. Green plants colonized the continents (4). On an Earth reshaped by life itself, they paved the way for the first amphibian, Ichthyostega (5).

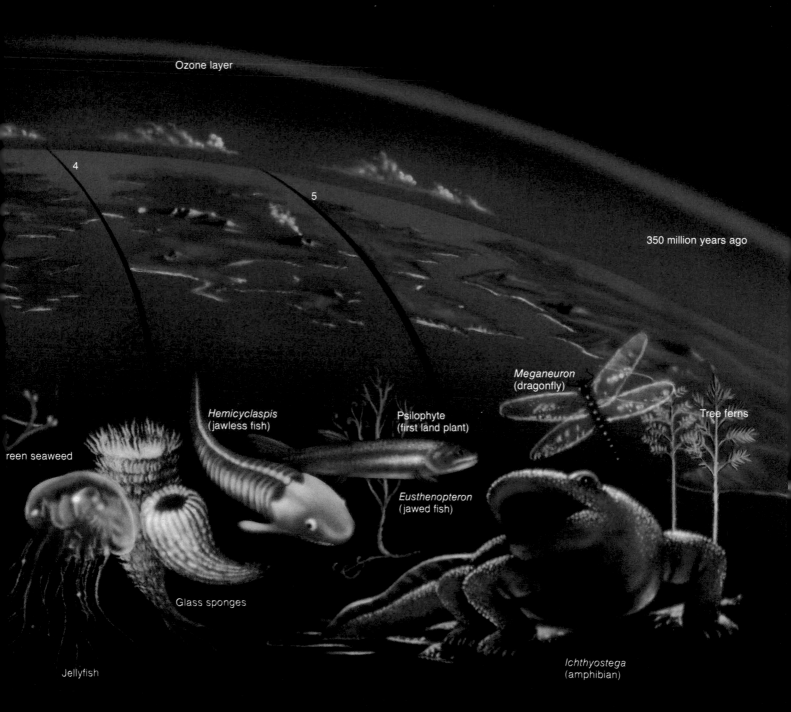

Ozone layer

4

5

350 million years ago

Meganeuron (dragonfly)

Tree ferns

Hemicyclaspis (jawless fish)

Psilophyte (first land plant)

reen seaweed

Eusthenopteron (jawed fish)

Glass sponges

Jellyfish

Ichthyostega (amphibian)

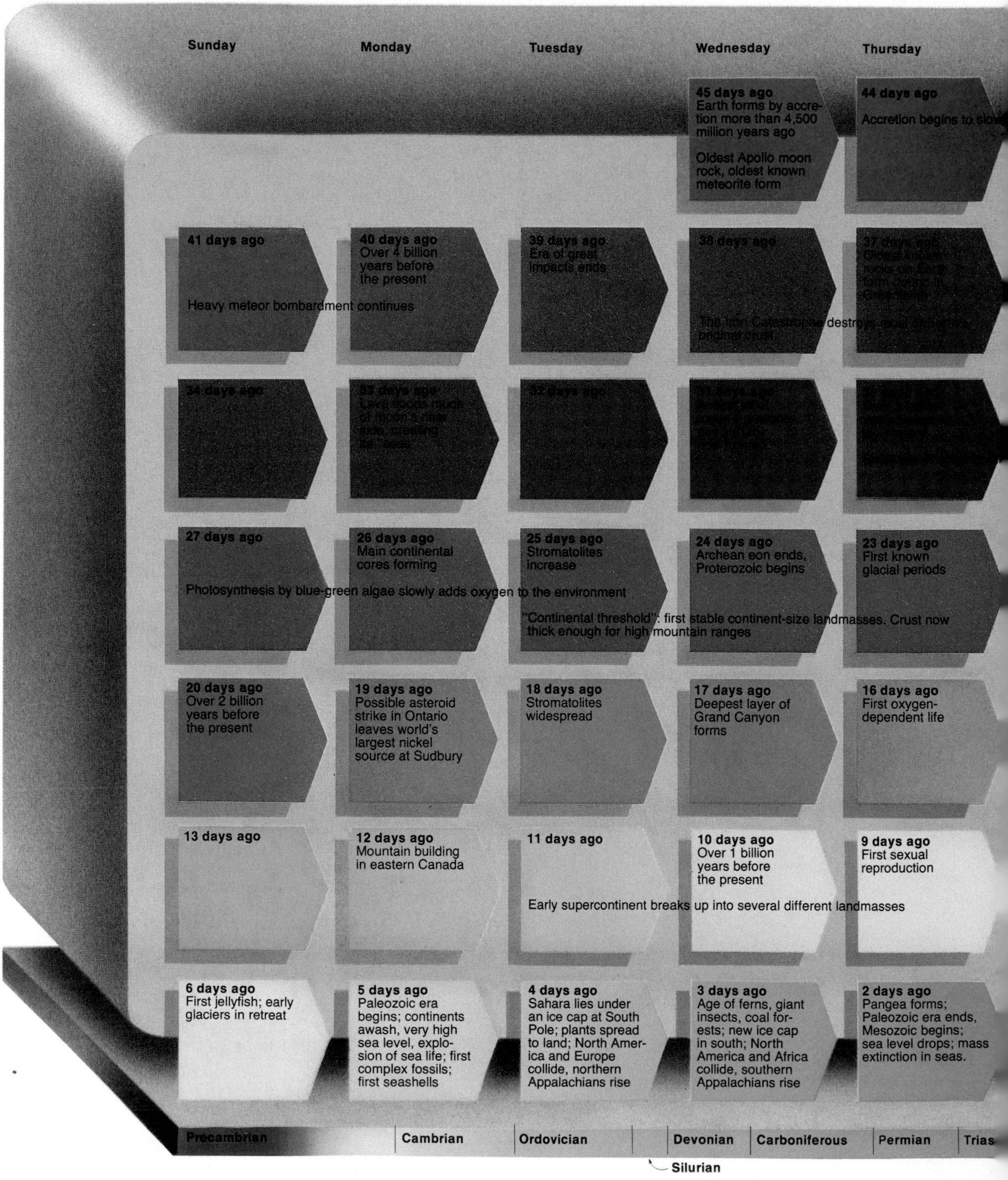

26

Sunday	Monday	Tuesday	Wednesday	Thursday
			45 days ago Earth forms by accretion more than 4,500 million years ago. Oldest Apollo moon rock, oldest known meteorite form	**44 days ago** Accretion begins to slow
41 days ago	**40 days ago** Over 4 billion years before the present	**39 days ago** Era of great impacts ends	**38 days ago**	**37 days ago** Oldest known rocks on Earth form (3800 m.y.) Greenland
34 days ago	**33 days ago** Lava floods much of moon's near side, creating its "seas"	**32 days ago**	**31 days ago**	**30 days ago**
27 days ago	**26 days ago** Main continental cores forming	**25 days ago** Stromatolites increase	**24 days ago** Archean eon ends, Proterozoic begins	**23 days ago** First known glacial periods
20 days ago Over 2 billion years before the present	**19 days ago** Possible asteroid strike in Ontario leaves world's largest nickel source at Sudbury	**18 days ago** Stromatolites widespread	**17 days ago** Deepest layer of Grand Canyon forms	**16 days ago** First oxygen-dependent life
13 days ago	**12 days ago** Mountain building in eastern Canada	**11 days ago**	**10 days ago** Over 1 billion years before the present	**9 days ago** First sexual reproduction
6 days ago First jellyfish; early glaciers in retreat	**5 days ago** Paleozoic era begins; continents awash, very high sea level, explosion of sea life; first complex fossils; first seashells	**4 days ago** Sahara lies under an ice cap at South Pole; plants spread to land; North America and Europe collide, northern Appalachians rise	**3 days ago** Age of ferns, giant insects, coal forests; new ice cap in south; North America and Africa collide, southern Appalachians rise	**2 days ago** Pangea forms; Paleozoic era ends, Mesozoic begins; sea level drops; mass extinction in seas.

Heavy meteor bombardment continues

The Iron Catastrophe destroys most of Earth's original crust

Photosynthesis by blue-green algae slowly adds oxygen to the environment

"Continental threshold": first stable continent-size landmasses. Crust now thick enough for high mountain ranges

Early supercontinent breaks up into several different landmasses

| Precambrian | | Cambrian | Ordovician | | Devonian | Carboniferous | Permian | Trias |

Silurian

46 Days of the Earth, at 100 Million Years a Day

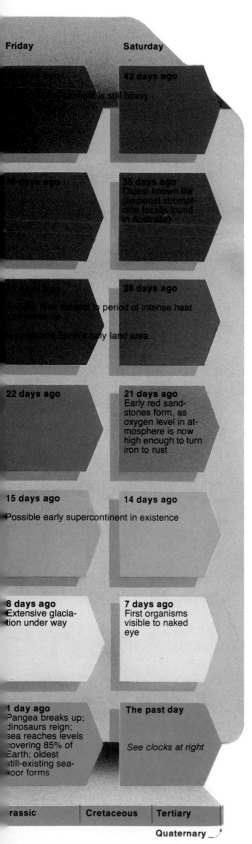

Friday

Saturday

42 days ago
...bardment is still heavy

35 days ago

35 days ago
Oldest known life (bacterial stromatolite fossils found in Australia)

28 days ago
...subjected to period of intense heat

...Earth's only land area

22 days ago

21 days ago
Early red sandstones form, as oxygen level in atmosphere is now high enough to turn iron to rust

15 days ago
Possible early supercontinent in existence

14 days ago

8 days ago
Extensive glaciation under way

7 days ago
First organisms visible to naked eye

1 day ago
Pangea breaks up; dinosaurs reign; sea reaches levels covering 85% of Earth; oldest still-existing seafloor forms

The past day

See clocks at right

...rassic | Cretaceous | Tertiary

Quaternary

For most of us, there is not much difference between 2 million years and 200 million. The numbers just blur into "a very long time."

But if we imagine a calendar on which each day represents 100 million years, our minds can begin to grasp the eons. On such a scale the Earth, born about 4.6 billion years ago, would be 46 days old. A mere million years passes in less than 15 minutes, a thousand years in just under a second.

On this calendar of the Earth, the end of the last day stands for the present. If you were to read, say, that a mountain range formed 250 million years ago, that would be at about noon, two days ago.

For the first few days of the calendar, great meteorite impacts continued as accretion wound down. The Iron Catastrophe took place a bit over five weeks ago. Then microscopic life appeared. As oxygen built up in the atmosphere, beginning about three weeks ago, pieces of continental crust drifted about, several ice ages came and went, and sea levels rose and fell.

Most events we know of occur in the last week, for Earth's churning crust has buried earlier clues, and the soft bodies of early life left few fossils. If the present time is midnight, Saturday night, then the Phanerozoic eon we live in opened last Monday morning—570 million years ago—when the Cambrian period began. The calendar's base aligns the geologic periods with the last six days. "Precambrian" refers to all time before then.

Things happened fast in this last week. Plant life appeared on land on Tuesday evening; insects and amphibians by Wednesday. On Thursday the continents drifted together into the supercontinent of Pangea; by evening the first dinosaurs dwelt there. Pangea lasted only a day, its parts dispersing by Friday afternoon. The dinosaurs' reign ended suddenly, at about 8 a.m. today. Mammals arose. A new glacial period—the one we call the Ice Age—began at 11:30 p.m. At nine seconds to midnight the last ice sheets retreated. The Normans invaded England about a second ago, and men landed on the moon, seeking rocks older than Earth's, in the last hundredth of a second.

The past day (present time is midnight)

8 a.m.:	2:15 p.m.:	7:00 p.m.:	10:10 p.m.:	11:45 p.m.:	11:59:50 p.m.:
Europe, Greenland separate; dinosaurs wiped out in wave of mysterious extinctions; Mesozoic era ends, Cenozoic begins	Australia drifting north from Antarctica; India collides with Asia	Red Sea opens; Midway Island is youngest and southernmost of Hawaiian chain; first great grasslands spread	Gulf of California opening; giant lake occupies much of the Zaïre (Congo) Basin; Africa closing Strait of Gibraltar	Panama has linked the Americas, spelling death for many species; hominids roam Africa and Eurasia; Ice Age under way	First towns; agriculture. At two seconds to midnight, Jesus of Nazareth born; at two-tenths of a second, the Industrial Revolution begins

The Bullard Fit—notching South and North America into Africa and Europe—used computers to verify Alfred Wegener's disputed concept of drifting continents. In 1965, Sir Edward Bullard and his Cambridge colleagues found the best fit at a depth of 1,000 m along most of the continental shelves.

Alaska's calving Columbia Glacier illustrates the retreat of huge Ice Age glaciers. Like an unloaded ship rising in the water, continental crust rises with loss of a heavy ice cap. Scandinavia uplifts about a meter every century. This principle of equilibrium, called isostasy, led Wegener to ask: If continents move up and down, why not sideways?

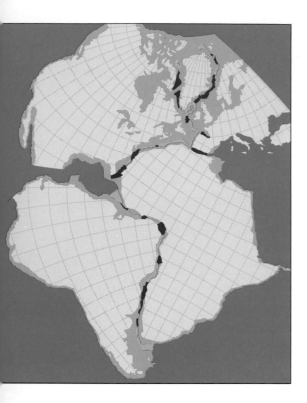

■ Overlap

■ Gap

■ Continental shelf

Eppur si muove! — But it does move!
Attributed to Galileo Galilei

As early as 1620 English philosopher Francis Bacon had commented that the coastlines of western Africa and eastern South America seemed to match. Benjamin Franklin 150 years later wondered whether our planet had a spherical core surrounded by fluids that buoyed up its rocky shell. Yet when I was in grade school in the 1950s, the books we read on geography and science either made no mention of continental drift or they cited it only as a curious, discredited theory. In those days a college professor who taught that the continents moved risked his academic reputation.

Now most scientists are sure the continents do move. That realization has revolutionized our picture of Earth's past and present. What happened to change the basic understanding of our planet? And why did it take so long?

Conflicting clues were already accumulating in the 19th and early 20th centuries. Paleontologists, for example, had found fossils of *Mesosaurus,* a Permian reptile, on both sides of the South Atlantic Ocean. *Mesosaurus* was not thought to be a great swimmer, certainly not able to cross an ocean. So geologists proposed that a great land bridge once connected the two continents.

In the tropics other scientists found sand and gravel left by ancient glaciers. Rainy regions yielded prehistoric desert sands; the arctic revealed ancient coal forests. In the early 1900s a British expedition discovered plant fossils only 400 miles from the South Pole. All sug-

gested tremendous shifts in climate.

In 1912 Alfred Wegener, a German meteorologist, amassed compelling arguments that attributed the confusing observations to the movement of the continents themselves. He based his idea on the out-of-place fossils and climates, on the fit of continental coastlines, and on new geologic facts scientists had collected. On each side of the Atlantic they had mapped rocks of various ages and compositions, as well as canyons, mountain ranges, and other structures. When they fitted the coastlines together, the geology matched too.

First called Continental Displacement and later Continental Drift, Wegener's thesis said that the continents had once been joined in a single landmass called Pangea, "all lands." When Pangea split up, its parts dispersed, eventually reaching the positions of the continents today. Flora and fauna, isolated on broken-off landmasses at new latitudes with different climates, followed differing evolutionary lines. Thus drift could explain everything: the fit of the coastlines, the puzzling fossil record, and the anomalous ancient climates.

The debate —
Wegener challenged
But not so fast. The same schools in Britain that had given science Hutton, Smith, and Lyell had also produced a prominent geophysicist named Harold Jeffreys, who challenged Wegener and his followers.

Jeffreys attacked Wegener's theory at its most vulnerable point, by asking *how* the continents could drift. How could a solid, stone continent plow its way through the

Early echo-sounding devices, forerunners of modern sonar aboard the ship below, profiled scores of flat-topped seamounts in the 1940s. Called guyots, they were later found to be remnants of volcanic islands, flattened by surf action. But how did their flat tops get so far below sea level? The theory of seafloor spreading, proposed in 1960, solved the guyot puzzle. The moving seafloor— contracting and sinking as it went —carried a volcanic island away from its birthplace into deepening water. After pounding waves leveled off the island, continued seafloor motion carried it deeper into the abyss. The result: a guyot.

solid, stone ocean floor? Wegener's theory proposed what had happened, but not how or why. It seemed to defy physics.

The attack prompted poorly supported rejoinders. Wegener thought the American continents were being propelled westward by tidal forces in the crust. Jeffreys replied that if these forces were strong enough to move continents, they were strong enough to halt the eastward rotation of the globe. Wegener advanced the idea that Earth's rotation created "flight from the poles," *Polflucht,* which had driven India from the Antarctic into Asia and crunched up the Himalayas. With simple calculations, Jeffreys demonstrated that such forces could not move continents— Earth's crust was too strong.

Theories about geophysics from a meteorologist found little acceptance among geologists, who questioned Wegener's credentials and claimed he selected only those facts that supported his hypothesis. In 1930 Wegener went off to Greenland, where he had done meteorological field work twice before. He died there in a blizzard. But the controversy he had raised lived on, polarizing earth science into "mobilists" and "fixists."

Survey ships
and fossil compasses

The debate between the two groups continued as technology and earth science advanced and curiosity about the ocean floor increased. Survey ships plumbing the depths for transatlantic telegraph cables in the 19th century had revealed mountainous terrain and shallower water in the middle of the ocean.

Scientists added to the picture after World War I with the echo sounder, a primitive sonar that bounced sound waves off the ocean floor and measured how long the echo took to return. It disclosed, vaguely, the topography of the abyssal depths.

Later, research ships employed seismic explosions that measured the density of rock layers by timing the shock waves through them. In 1947 an American team aboard the research ship *Atlantis,* headed by Maurice Ewing and Bruce C. Heezen, found that sediment depth on the Atlantic floor was not nearly what long ages of accumulation should have yielded. Why so thin?

A different kind of surprise came in the 1950s and 1960s when accumulated data from oceanographic centers and ship trackings showed that a mountain range 74,000 kilometers (46,000 mi) long snaked through all the seven seas—the Mid-Ocean Ridge.

The key to the mystery of how our planet works, however, lay in another mystery, the phenomenon of magnetism.

Some twenty-five hundred years ago the Greeks discovered the magnetic properties of the lodestone. It wasn't until the 1300s that the Chinese floated a lodestone on a piece of wood in water and invented the compass. It took another 300 years before British physicist William Gilbert figured out why the compass needle aligned itself north and south. He proposed that the whole Earth was one big magnet whose force acted on smaller magnets. Scientists now suspect that Earth's fluid outer core, spinning like the dynamo in a power plant, sets up the magnetic field.

Not only lodestones have magnetic properties. When molten rock, or magma, flows to the surface, grains of iron oxide in the liquid stone line up with the magnetic poles like compass needles. As the magma hardens, the grains freeze in position, creating a record of where the wandering magnetic poles are when the rock forms. The magnetism in this rock is feeble and difficult to read. But in 1953, British physicist P.M.S. Blackett developed a supersensitive device able to detect the weakest magnetic fields —the astatic magnetometer. Scientists used it to read magnetic polarity in the rock.

The magnetism in rocks had many stories to tell. One came out of Australia in 1971. A group investigating an aboriginal campsite 30,000 years old found that the heat from the fire had allowed iron particles in the stones to realign with Earth's magnetic field at that time. They pointed south! Magnetic north 30 millennia ago must have been somewhere in the Antarctic. The discovery not only confirmed the theory that the poles actually reverse their magnetic fields from time to time but showed that they have done so recently.

We don't know why, but major reversals occur about every half million years, and shorter flips, "magnetic events," last from a few thousand to 200,000 years. Some unknown motion in the fluid of the outer core may cause the reversals, but whatever the reason, their patterns give scientists a new yardstick for gauging geologic age.

Another piece of the time puzzle had come from World War II. Allied shipping routes were under

Water gushes on Challenger's deck as the drill pipe is "tripped"— pulled up—disconnected, and stored. Oil industry technology designed the ship, operated by the National Science Foundation, but this vessel's riches lie in the plastic tube crewmen remove from a metal pipe (left). Cut into three-foot cylinders and sliced lengthwise (below), its compressed contents— fossils, lava, and sediments— confirmed seafloor spreading: Cores from either side of the Mid-Ocean Ridge became progressively older away from the Ridge; they showed that the Atlantic grows nearly an inch wider a year.

The cores revealed other seabed history: No rock or sediment in the oceans has been found older than 200 million years; the oldest ocean floor is in the Pacific's northwest corner; and the Mediterranean Sea has dried up—perhaps more than once—in the past 12 million years.

constant attack by German U-boats. To counter this menace, the Allies developed the Magnetic Airborne Detector (MAD), which could locate the steel hull of a submerged submarine when lowered into the water from a blimp or ship.

Oceanographers get MAD

After the war, oceanographers modified the MAD and began towing it behind ships to identify magnetic, mineral-rich rocks. They soon detected odd magnetic patterns in the seafloor. Long zebra stripes of normal and reversed rock alternated in a symmetrical pattern on each side of the axis of the Mid-Ocean Ridge.

Scientists had been exploring these mountains for a decade or more, trying to determine their origin, when, in 1960, Harry H. Hess, a Princeton geologist, offered an explanation. As a Navy commander in the Pacific he had made depth soundings and discovered whole strings of submarine volcanoes in progressive stages of erosion. He also knew about the thin sedimentary layers Ewing and Heezen had found in the Atlantic; to him they meant the ocean floors were young. And he believed that magma circulating in convection currents under the crust moved the ocean floors and continents.

From all this Hess reasoned as follows: Molten rock must be oozing up out of Earth's interior under the sea, creating the mountains of the Mid-Ocean Ridge and spilling out new seafloor that spreads away to each side. The seafloor carries along strings of submarine volcanoes which erode and sink deeper under the sea with the cooling crust. The spreading seafloor eventually descends into the mantle, forming ocean trenches as it goes.

Since Hess could not prove any of this to a scientific community still hostile to the notion of a movable crust, he offered his theory under the disarming and undogmatic label of "geopoetry."

Magnetic stripes — clues to the mystery

Among the first to support Hess were Cambridge geologists Frederick J. Vine and Drummond H. Matthews, who puzzled over the magnetic stripes along the Mid-Ocean Ridge. The symmetry of the stripes could not be accidental. Earth was trying to tell them something. In their search for understanding, they found evidence that Hess's poetry was factual prose.

Readings from the deep showed in the early 1960s that molten rock did emerge along the axis of the Mid-Ocean Ridge. As the lava hardened, the miniature compasses within it locked into place, recording Earth's magnetic polarity at that moment. At or near the crest of the Ridge, where the upwelling was going on, rocks should be very young and have today's polarity. They were, and they did. North was north. To either side of the Ridge, rocks should be much older and have polarities reflecting the last magnetic reversal. They were, and they did. North was south.

Since half the reversed rock was on one side of the ridge and half on the other, and since both strips lay at about the same distance from the summit, the rocks probably once formed a single band, which had been ripped in two lengthwise. Half the rock went one way beneath the ocean and half the other.

If the magnetic evidence was true, then Hess had to be right: The floor of the ocean did spread. The Mid-Ocean Ridge marked the beginning of two giant conveyor belts that moved new crust from the Ridge toward each side.

Some skeptics, however, demanded further proof. They got it from an unexpected source.

In the decades following World War II, the industrial nations came to realize they were depleting the world's oil reserves on land. The search for offshore oil was on, a quest that required equipment capable of working in ever increasing depths. Oil companies mounted some rigs on platforms and others on ships that were modified to carry four or five miles of drill pipe.

So earth scientists took their ideas about seafloor spreading to the oil industry, which designed the *Glomar Challenger,* the first drilling ship built to work in the deep waters of the ocean basins.

In 1968 and 1969 *Challenger* undertook a historic traverse between the continents of Africa and South America, on whose matching coastlines Wegener had partially based his case. Now scientists could probe the seafloor in between, exploring where Wegener could not.

Drilling brought up core samples of fossils that revealed a stage-by-stage increase in ocean-floor age as distance from the Mid-Ocean Ridge increased. With the samples Hess's theory came a step nearer vindication. As refined by other earth scientists, the theory now gaining acceptance was this: The supercontinent of Pangea had once existed,

Earth's magnetic field helped prove that seafloors spread. Our globe's polarity has flip-flopped repeatedly over millions of years, from normal to reversed (when the north end of a compass needle points south). When fresh lava at mid-ocean ridges cooled, magnetic particles within the lava lined up with the field like tiny compasses, *recording the direction of magnetic north at the time. In the 1950s and 1960s, magnetometers revealed a symmetrical pattern in seafloor rock: alternating bands of normal (yellow) and reversed polarity. Each pair of bands had formed in the center, only to be torn in two as the seafloor spread.*

and it had split up. Africa and South America did move apart. The continents did not have to plow through stone seafloors as Wegener had helplessly supposed; the seafloors themselves moved, taking the continents with them.

But the puzzle needed a final piece. Seafloor spreading explained how the Ridge was creating crust, but not where the crust was going. Hess had said it was disappearing down oceanic trenches.

Possible confirmation, it was thought, might come from seismology, the science of earthquakes. It was known that earthquakes and volcanic eruptions do not occur randomly across the planet but in narrow seismic zones, which include the Mid-Ocean Ridge and the deep ocean trenches. One such zone, the "Ring of Fire," encircles the Pacific with a system of trenches and parallel volcanic ranges.

Now, new seismological technology had emerged, not from war, but from an attempt to control the weapons of war—the 1963 treaty to ban nuclear weapon testing everywhere except under the ground. To monitor underground tests, more than 125 seismographic stations were set up around the world. It was in the role of watchdog that this network over the years also provided information that helped confirm the final part of Hess's theory. The seismographs, so sensitive they could tell a bomb blast from an earthquake, gave scientists a view of earthquakes that showed crust descending at the trenches.

The two major discoveries—of the extent of the Mid-Ocean Ridge and of seafloor spreading—were what led to the revolution now

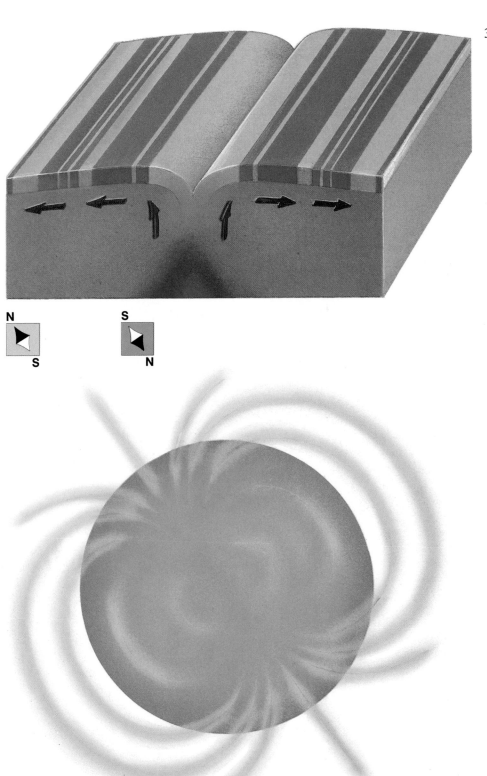

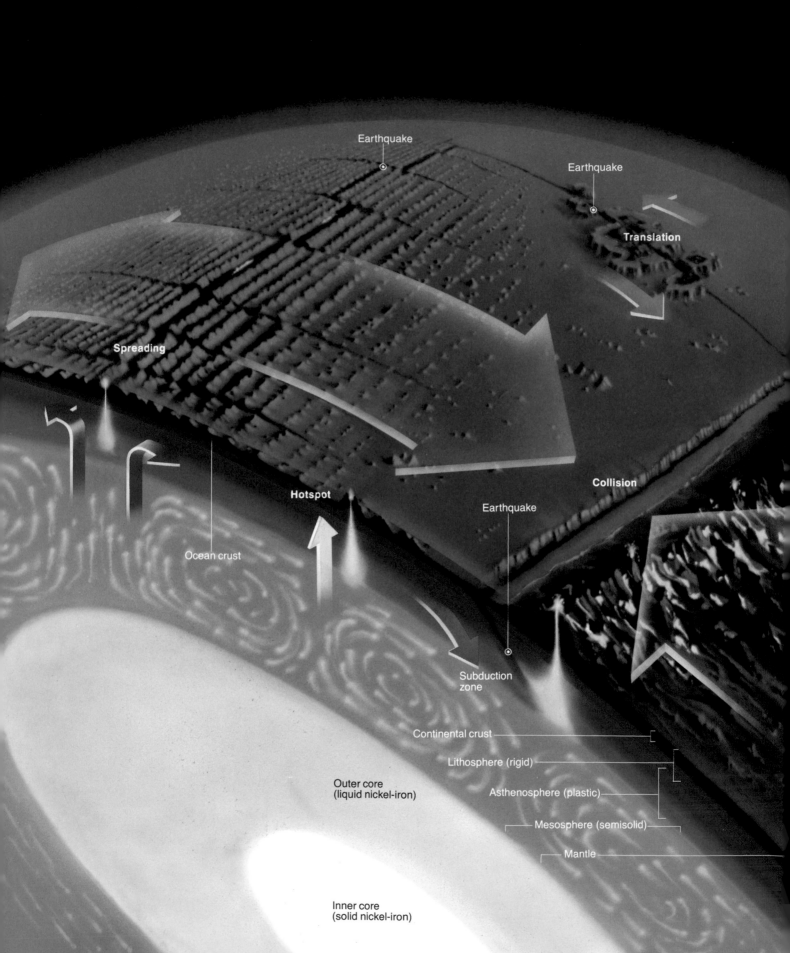

Earthquake

Earthquake

Translation

Spreading

Collision

Hotspot

Ocean crust

Earthquake

Subduction
zone

Continental crust

Lithosphere (rigid)

Outer core
(liquid nickel-iron)

Asthenosphere (plastic)

Mesosphere (semisolid)

Mantle

Inner core
(solid nickel-iron)

Interior heat churns our living planet to self-renewal and drives its crustal plates. Earthquakes occur at their boundaries. Where an ocean ridge marks a spreading zone, upwelling molten rock makes new seafloor. Magma can also break through a plate at a hotspot, building volcanoes that plate motion carries away. When plates collide, continental crust crumples into mountains and the seafloor dives, creating a deep trench. In a quake-prone subduction zone, the plunging plate remelts. Its descent generates magma that creates and fuels volcanoes above. In zones of translation, passing plates slip and snag along fault lines, jostling the crust with more earthquakes.

called plate tectonics. Tekton was the name of the carpenter in the *Iliad,* and "tectonic" has come to mean "construction." Tectonics is the branch of geology that studies the structure of the globe itself.

Plate tectonics not only vindicated Wegener, it transformed geology as profoundly as the theories of evolution and relativity transformed biology and physics. In addition to revising earth-science textbooks, the new theory would cast light on seemingly unrelated questions: why climates shift, where metals come from, why some species die off and others change.

How plate tectonics works
The period following *Glomar Challenger's* early discoveries was busy and exciting. I was a graduate student then, and my doctoral thesis, like many of the time, sought to fit aspects of continental geology in with plate tectonics, a theory based on unseen events beneath the sea.

According to this theory, the seismic zones detailed in the 1960s outline large slabs that form the Earth's outer skin—the plates (see map, pages 58-9). Though they may measure several thousand kilometers across, plates are only 50 to 150 kilometers thick—about as thick in proportion to Earth as an eggshell to an egg. A few big plates and many smaller ones make up the rigid *lithosphere,* which floats on a partly molten layer of rock known as the *asthenosphere.* The plates move—most of them away from ridges and toward trenches.

The movement of the plates makes different things happen at their edges. Where two plates move apart, as they do in the Atlantic, for example, their motion creates new crust and a ridge forms.

Where plates collide, as they do at the perimeter of the Pacific Ocean, they usually destroy crust. An ocean trench marks the region where one plate bends downward—*subducts*—beneath the edge of another. The pressure of this motion often wrinkles and buckles the upper plate, building mountain ranges like the Andes along its edge. The subducting plate angles down into the Earth's interior, where it makes the overlying rock melt and, deeper down, gradually melts itself. The magma rises, fueling volcanoes in the mountain range above.

But if both colliding plates carry continental crust, as occurs where India meets the Eurasian Plate, neither plate subducts completely. Instead, the impact shoves the crust skyward, building great mountain chains like the Himalayas.

Where plates slide sideways past one another, as along California's San Andreas Fault, their motion causes earthquakes but neither creates crust nor destroys it.

In some places a fourth activity, possibly a plume of rising magma within the mantle, creates a *hotspot* that can melt right through the plate above it. The result is a line of volcanoes like the Hawaiian Islands, built one after another as the plate moves across the hotspot.

We can see the effects of plate motion on Earth's surface, but the question of what makes the plates move—the same issue that stymied Wegener—remains a mystery.

To try to solve it, geophysicists investigate the plates and the regions beneath them in the upper mantle. In the Pacific Ocean Basin, they believe, the plates may be partly self-propelled: As the leading edge of a plate subducts, its weight pulls the rest of the plate toward the trench, the way a tablecloth slides from the table once enough of it is hanging off the edge.

But that theory cannot explain plate motion under the Atlantic: There is very little subduction at that ocean's edges. The force driving the Atlantic's plates, as well as contributing to Pacific plate motion, may be convection within Earth's mantle. Geophysicists believe that hot rock in the mantle moves like air in a room warmed by a radiator. The air rises to the ceiling, flows horizontally there, and then returns to the floor again as it cools. Similarly, hot mantle material rises—perhaps even causing hotspots—and moves horizontally beneath the rigid lithosphere, propelling the plates as it circulates.

Young seafloors, ancient continents
Geologists' explorations since early in this century have turned up an unexpected age pattern in the Earth's crust. Scientists have found rocks some 3.8 billion years old on land, but the oldest rock recovered from the ocean floor is less than 200 million years old. Earth's seabeds must be recycling. Why?

Undersea topography holds the answer. Where two seafloor plates move apart, a ridge forms—not because of buckling but because of thermal expansion of the rock. The ridge is a welt, swollen with heat, and it lies atop upwelling material in a similarly swollen part of the upper mantle, or asthenosphere.

As the plates shift away from the ridge crest, the new crust at their

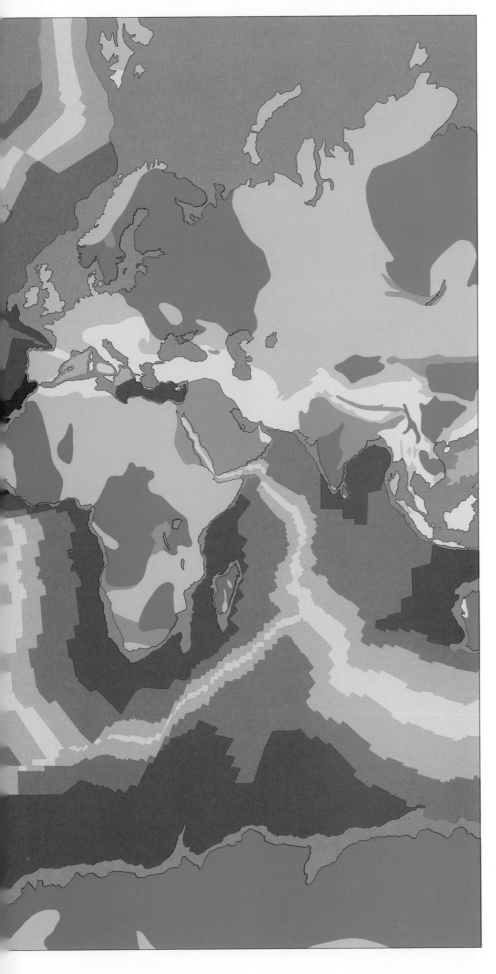

Age of Earth's crust shows striking patterns revealed by magnetic and fossil analysis and by radioactive dating methods. Except for their mountainous margins (light green), the continents, whose dark green tones designate older rock, stand out from younger ocean floor. The continental margins often mark plate boundaries (see pages 58-9) where collision has reshaped the rock through buckling and volcanism. As clear as tree rings, the parallel stripes of color on the young ocean floors show a history of seafloor spreading that continues today. But it is the continents, whose age pattern is less regular, that hold the clues to our planet's deep past.

The reason is that continental crust is relatively light and buoyant. In a plate collision, it tends to stay on top. Pieces of continental crust riding a plate into a subduction zone don't descend into the mantle. Instead, they may fuse with a craton, *or continental core, on the overriding plate. Since continents can grow by scraping such fragments from descending plates, continental* shields—*where the old craton is exposed on the surface, as in Canada—lie surrounded by younger rock.*

Age in millions of years

Continental crust

	0 to 250
	250 to 800
	800 to 1,700
	1,700 to 3,800
	Continental shelf

Oceanic crust

	0 to 9
	9 to 35
	35 to 80
	80 to 140
	140 to 180

Eight freeze-frames from a half-billion-year movie trace Earth's moving continents through the past. Red lines show areas that will become landmasses of today—some easy to recognize, like Africa, embedded upside-down in ancient Gondwana (first map). Others are pieces of continents-to-be, named to indicate their future roles: Baltica, for example, will become northern Europe; Laurentia (named for Canada's Laurentian Highlands), northeastern North America. Fluctuating sea levels in some eras often flooded continental crust. Climatic and magnetic evidence gives scientists a firm idea of the continents' north-south movements, but east-west positions before the breakup of Pangea (about 200 million years ago) are less sure. From then on, spreading seafloors left a more reliable record, but paleogeographers still debate each other's conclusions.

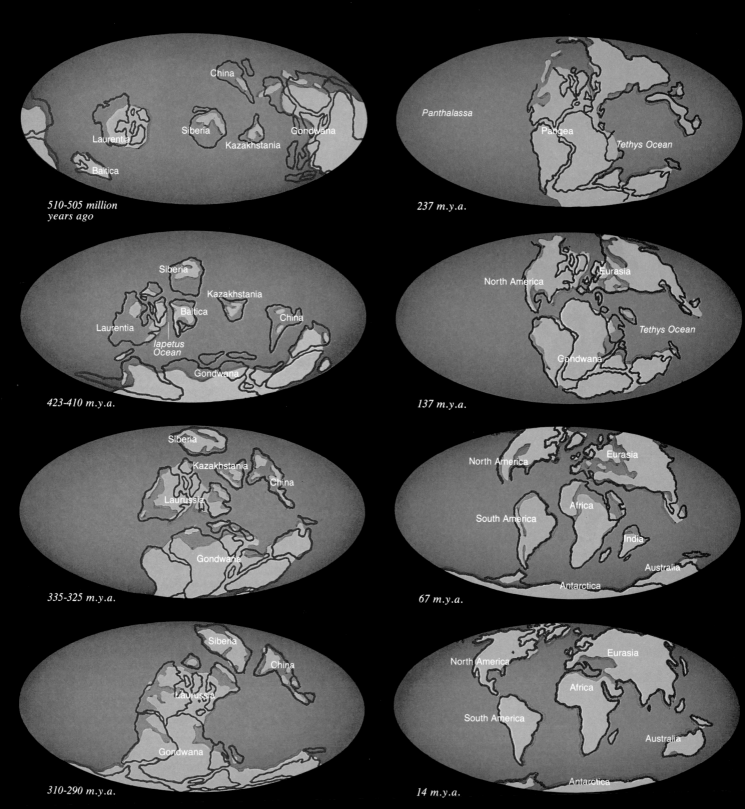

510-505 million years ago

423-410 m.y.a.

335-325 m.y.a.

310-290 m.y.a.

237 m.y.a.

137 m.y.a.

67 m.y.a.

14 m.y.a.

Buried in the chalky bottom of a Tethys lagoon some 50 million years ago, this tall-finned Eocene fish left its imprint and its skeleton engraved in mud that later became hard limestone. Colliding plates raised part of the former seabed to form Monte Bolca in northern Italy, where scientists split a rock to find the fossil.

edges gradually cools and contracts. As the rock contracts, it grows denser and sinks. That's why, when a high-floating slab of young oceanic crust collides with an older slab, the younger one remains on top. The older rock, forced to dive beneath the younger, melts back into the mantle. And so the ocean floor stays young.

But why haven't the foundations of the continents been recycled in the same way? How have they survived plate collisions over billions of years? Again, density is the answer. Seafloors are denser than continents, thanks to their composition. Continental crust is mostly granite, and granite is 90 percent lightweight quartz and feldspar. Oceanic crust, on the other hand, is iron-rich basalt, a very heavy rock. So when plate motion brings dense seafloor into collision with more buoyant continental crust, the continental crust floats on top, and it is the seafloor crust that subducts.

The time detectives

All these motions and collisions add up to tremendous changes in the face of the Earth over time. What was the ancient world like? Just as forensic scientists can reconstruct a dead person's face from clues on a fleshless skull, paleogeographers can reconstruct the past faces of the Earth from scattered clues—rock magnetism, the location of climate-sensitive deposits like coal and bauxite, and oxygen isotope ratios in fossil shells.

At the University of Chicago, a team headed by Alfred Ziegler and Christopher Scotese is working on a paleogeographic atlas showing Earth's changing surface over the last 600 million years. It will include not only the location of the continents but also shorelines and mountains, plant and animal life, rock types and geologic formations, climates and ocean currents. "Plate tectonics is the theme that unifies the atlas," Scotese says. "It gives us a way to fit together facts from all branches of the earth sciences."

What Ziegler and Scotese are doing, in effect, is building us a time machine like the one imagined by H. G. Wells. In it we will travel millions of years in the blink of an eye and see where the continents used to be. We will discover that their wandering did not just rearrange the map; plate movement also helped ice ages come and go, seas rise and fall, and deserts spread and retreat. It even shaped life itself.

The grand tour of all time

Let's start our trip in the Archean eon, some 3.8 billion years back, at the age of Earth's oldest rocks. The partly molten planet is in the throes of the Iron Catastrophe, releasing heat into space. Huge clusters of volcanoes have built up on the surface, spewing gases and lava.

As we fly through the millions of years, changes on the Earth's surface tell us what is happening below: Separate streams of upwelling molten rock in the mantle come together, forming great currents of rising magma. They resemble the spreading ridges active on Earth today, and may even have evolved into them. Elsewhere, buoyant continental crust accumulates above "cold sinks," regions where dense, cooled material descends again.

Living leftovers from an ancient Precambrian heyday, stromatolites crowd an Australian tidal pool, safe from wave and predator. They form when sand, silt, or mud covers mats of blue-green algae; the algae cement down the sediment, then grow over it. Fossil stromatolites 3.5 billion years old are the earliest non-microscopic evidence of life.

44

Archean rock that has survived into our own time reveals the heat and violence that forged Earth's crust three billion years ago. Most of the granitelike rock has been repeatedly baked and compressed, fantastically folded and deformed— far more extensively than today's tectonic processes could accomplish. Yet Archaean crust also contains areas of sedimentary and volcanic rock that somehow escaped alteration, as well as plentiful batholiths—huge domes of once molten rock that rose into the crust from below like giant balloons.

A visit to this Archean terrain in its youth puts us on an utterly alien Earth. The total area of the continents is perhaps one-eighth of their extent in our own time. The ground is barren and the seas hold only microscopic life. The atmosphere is unbreathable—no free oxygen and too much carbon dioxide—and the sun's ultraviolet light blasts down unfiltered by an ozone layer. The days are less than 24 hours long, and the moon seems huge, for it orbits Earth far closer than today's 384,400-kilometer distance.

When we travel on to the beginning of the Proterozoic eon, 2.5 billion years before the present— "BP," as geologists say—we see that our planet has emerged from the tumultuous Archean into a time of comparative calm. The same tectonic forces as in our day are at work. The Proterozoic marks the "continental threshold," when the continents have nearly reached the size typical of the modern world.

Otherwise, the Proterozoic seems little different from the Archean. But blue-green algae—actually a type of bacteria—are photosynthe-

sizing in the oceans, pumping oxygen into the atmosphere. The continents raft together into a supercontinent, a predecessor of Wegener's Pangea, that will not break up until about one billion years BP.

If we could drill deep into the Proterozoic Earth, we would find huge bodies of magma cooling undisturbed in the crust—evidence of a long period of geologic peace. Metal-rich layers form in them as minerals progressively crystallize and settle out of the melt. We know the richest of these magma bodies as the Bushveld Complex—the source of South Africa's wealth of chromium, platinum, and nickel.

Continuing toward the present, we enter our own eon, the Phanerozoic—"visible life." Its first period, the Cambrian, begins only 570 million years BP. Most landmasses now lie in a garland around the Equator. The largest, called Gondwana, contains the nuclei of Africa, South America, Antarctica, Australia, and India. There has been a tremendous explosion of life in the sea since our last stop, and many new life forms will leave fossils that give us clues about climate and continental movement.

Hard fossils, hard facts

Why this sudden emergence of varied life forms? Greater abundance of free oxygen is probably the main reason, but plate motion also contributes. There is extensive seafloor spreading in the Cambrian, and when spreading rates are high, growing mid-ocean ridges occupy more space in the seas. As the ridges swell, they displace water, raising sea levels worldwide and *(continued on page 54)*

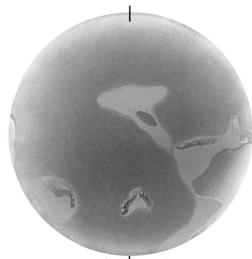

Dawn illuminates *the watery world of the early Cambrian, 540 million years ago. Vast tidal flats surround the equatorial island continents that compose most of Earth's land area. Warm shallow seas have flooded much of tropical Siberia (lower left island on globe), and both sea and shallows teem with life. In these favorable conditions, living things change rapidly; for the first time since life began, many ocean species develop protective shells of calcium carbonate or of chitin. These hard parts, like the exoskeletons of the trilobites at left, will leave visible fossils when the waves bury them under sediment. Broad-topped, layered stromatolites grow on the tidal flats, protected from algae-grazing mollusks by the high salt content of the sun-warmed water. The sun's lethal ultraviolet light still makes life on dry land impossible in the Cambrian period, but it does little damage to the stromatolites' blue-green algae. They are periodically protected by high tide, and, as a form of bacteria, are very resistant to radiation damage.*

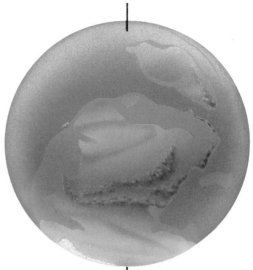

A morning thundershower *sweeps southwest along the Appalachian Mountains of a tropical Kentucky. These fresh, jagged peaks, nearly as tall as today's Rockies, are still rising here in the Carboniferous period, 300 million years ago. The force that raises them is the collision of Gondwana (south of mountains on globe) with Laurussia, made up of North America, Europe, and Greenland. Through these Kentucky marshes, lush with tree ferns, conifers, and giant club mosses, a river meanders toward a swampy bay. A dragonfly the size of a dinner plate swoops past. As glaciers advance and retreat near the South Pole, sea levels fall and rise, and flat coastlines shift back and forth. Layers of mud, silt, and sand eroded from the mountains cover and compress thick beds of dead vegetation. Over the millennia, the heat and pressure of deep burial will transform this organic matter into coal.*

In the north, an encroaching Siberia has already forced up the Urals as an island mountain chain.

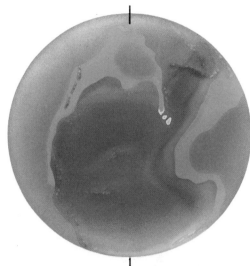

A sudden dust storm *routs a herd of* Plateosaurus *dinosaurs from a meager river in the red semidesert of what will become southern Germany. In the Triassic period, 240 to 205 million years ago, the supercontinent of Pangea includes almost all the world's land. Europe lies between the Ural Mountains and a shallow arm of ocean to the west (upper right on globe). Much of Pangea is flat and dusty, punctuated with worn mountains, sand dunes, and occasional oases. Oxidized iron gives its soils the brick red color characteristic of the Triassic—evidence of the warm, arid climate that prevails over most of the supercontinent. With only one continent and one ocean, regional differences among animals and plants are few. Ancestors of today's lizards and turtles, along with the first primitive mammals, share the late Triassic with dinosaurs like* Plateosaurus, *the 8-meter-long (26 ft), plant-eating forerunner of the great "thunder lizards" to come.*

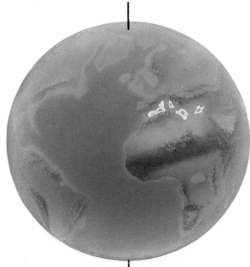

In the cool of a Miocene night *six
million years ago, a family of*
Gomphotheres, *four-tusked
ancestors of modern elephants,
migrates toward Europe across
the salt flats of a dried-up
Mediterranean Sea. A distant
volcano testifies to the violence of
Africa's northward advance on
Eurasia, still going on today. Here
in the Miocene, that continental
crunch has temporarily raised the
threshold between Mediterranean
and Atlantic and cut off the
inflow of Atlantic seawater. The
Mediterranean has evaporated,
leaving scattered brine lakes in a
basin 3,000 m (10,000 ft) deep. Salt
and other* evaporites *will layer its
floor 2,000 m thick. Without that
salt, the rest of the world's oceans
freeze more easily; the Antarctic
ice cap grows to one and a half
times today's size, and sea levels
fall. Cut off from the Indian
Ocean, the Tethys survives as a
much-reduced inland sea northeast
of the Mediterranean salt flats.
North and South America are
still separate, but close to their
present-day positions.*

Red kangaroos, marsupials of Australia, owe their shape to plate tectonics. Marsupials probably first appeared in South American forests when that continent was still joined with Australia and Antarctica—though only a few kinds survive on South America today. A marsupial jawbone found in Antarctica proves they lived there too, until ice engulfed them. Australia's drift northward saved its marsupials from the same fate, and its isolation kept out competing species. As the continent reached drier latitudes, forests gave way to grasslands, favoring the development of long-legged, grass-eating marsupial species: today's kangaroos.

inundating continental lowlands. In the Cambrian, as in our own time, tropical climates produce the greatest variety of species. The flooded margins of the equatorial continents provide vast warm shallow seas with many different ecological niches for new species.

We travel on from the Cambrian, watching continents shift like ice floes on a sea. Some 450 million years BP, in the Ordovician period, Europe and North America begin to close the Iapetus (proto-Atlantic) Ocean. Offshore island chains collide first, crumpling the continental shelf into peaks. We know part of their eroded remains as the Green Mountains of Vermont.

Ozone, plants, and coal

As we pass into the Silurian period, only 435 million years back, the ozone layer in the upper atmosphere has grown dense enough to block most of the ultraviolet radiation. Plants are colonizing the land, the first life to leave the protection of the oceans. In the Devonian, animals follow the plants, and late in the period, some 365 million years BP, the first trees—and Earth's first forests—appear in the lowlands. Primitive insects, scorpions, and spiders crawl among them, some falling prey to the first amphibians.

Our next stop, 300 million years BP, is in a steamy jungle. This is the middle Carboniferous, a period of great economic importance for the future. North America and Europe straddle the Equator. The African part of Gondwana is colliding with the North American part of Laurussia, creating the southern and central Appalachians.

Another region of Gondwana lies over the South Pole. As the size of the ice cap varies, world sea levels fluctuate, repeatedly flooding the continents. Sediment from the new mountains chokes the rivers, which build swampy deltas in inland seas. Rooted in dense mats of dead vegetation, forests flourish.

It's a perfect climate for the spread of amphibians—and for creating coal. When vegetation grows so fast that it buries dead layers before they can decay, and when the sediment of unstable river deltas helps seal the layers in, then deep subterranean heat and pressure will eventually turn this organic matter into coal. Some 20 percent of North America's coal reserves formed in Carboniferous times.

By the Permian, 290 million years BP, all the continents except China have joined into a single landmass: Pangea. Collision has forced the Appalachians to enormous heights, blocking rainfall to the northwest and turning the coal forests to desert and dry savanna. Many habitats change drastically. Similar species that once inhabited separate continents now share one landmass and must compete for survival.

Times are hard for land and sea creatures alike, and thousands of species disappear in the broadest extinction the world has known. But those that can adapt to the new conditions prosper. Reptiles proliferate, for example, because their eggs, unlike those of their amphibian ancestors, have shells and can survive far from bodies of water.

Our time trip now takes us into the Triassic period, 240 million years BP. Pangea is drifting north. The land is warm and glacier-free. The arid climate is nearly the same over most of the vast continent, and so too are the animals. Early dinosaurs and marine reptiles find conditions to their liking; on land, the first small mammals have appeared.

Pangea: the beginning of the end

As the Jurassic period begins, 205 million years BP, tectonic forces are at work beneath Pangea, straining the seams between its component landmasses. North America shows signs of splitting off. As New England pulls away from Morocco, rift valleys open in the crust, similar to those that crease Africa in our own time. They become a widening sea: the infant North Atlantic.

As we travel through the Jurassic period to the Cretaceous, 138 million years BP, we can see an arm of the Tethys Ocean growing westward across the tropics, dividing Pangea into Gondwana in the south and Laurasia in the north. Gondwana itself—at least 400 million years old—is also breaking up, as South America begins to separate from Africa. Many of the new fractures follow the lines of old joins, but not all, and not exactly: Boston, for example, is on a piece of crust that may have been part of Africa.

Rapidly spreading seafloors disperse the continents. By the late Cretaceous, sea levels rise almost 500 meters above today's height; only 15 percent of the globe's surface remains dry.

Stopping to sample the water of the Tethys, we find it incubator-warm, shielded by the continents from polar currents. For 30 million years, vast populations of plankton thrive in these equatorial waters. The plankton sink to the seafloor when they die. Like vegetation

Utah's Provo Canyon shows layers of limestone, sandstone, and shale laid down in a Carboniferous sea. The pressure of Pacific Ocean plates on North America's western edge gradually bent the rock some 100 million years ago. Crust and sediment scraped off those subducting plates have added new land to the continent.

turned to coal, their remnants too are transformed by heat and pressure—into petroleum. The rich oil fields of the Middle East and North Africa, with almost 60 percent of the world's proven reserves, tap the rock that formed at the Tethys Ocean's southern margin. The remains of Cretaceous plankton may be filling your gas tank today.

And suddenly at the end of the Cretaceous, the dinosaurs, lords of the Earth since the late Triassic, are gone. Perhaps a great shift in the world weather system erased them from the planet . . . or perhaps it was the effects of a monster meteorite impact 65 million years ago.

Scientists have found a thin, iridium-rich layer between Cretaceous and Tertiary rock all over the world. Because most of Earth's iridium sank in the Iron Catastrophe, the layer is good evidence that a large extraterrestrial object did crash into Earth at the end of the Cretaceous. The impact might have sent a blanket of dust into the atmosphere, blocking the sunlight on which plants depend and disrupting ecosystems everywhere. If the meteorite struck in the ocean, it would also have sent huge waves across the land and possibly heated the seas. But nothing has yet firmly linked the iridium layer to the dinosaurs' extinction, so the dinosaur mystery is far from solved.

With the dinosaurs gone, we enter the Cenozoic era, the Age of Mammals. By about 40 million years BP, a great convergence of continents is building mountains from Spain to Burma. Africa and Arabia drive northward, all but eliminating the Tethys; Italy rams Europe, raising peaks that preced-

ed today's Alps. The movements of a swarm of microplates create the Mediterranean seafloor.

Immigrants and a special newcomer
Pangea's breakup has gradually isolated its animal populations on separate continents, where they develop in separate ways. Each continent soon becomes home to quite different families of life. As North and South America drift toward the positions they hold in our time, the northern continent harbors mastodons, tapirs, early camels, and placental carnivores. South America, an island continent broken from Gondwana, carries armadillos, giant sloths, and a host of marsupials.

About three million years BP, colliding ocean floors raise the Isthmus of Panama between the Americas. As the land bridge appears, animals migrate across it in both directions. The armadillo succeeds in its new northern range, but many mammals, especially South American ones, such as the *Thylacosmilus*, a jaguar-size saber-toothed marsupial, cannot survive the competition. They die out—a mass extinction caused by plate motion.

About a million years later, a new creature appears in the rift valleys of Africa. Over the next two million years the descendants of *Homo habilis* will take control of their world and their destiny; no other life form since blue-green algae has so transformed the Earth. Yet they—we—are just beginning to understand Earth's powerful, patient forces. Indeed, scientists today have returned to explore those very same rift valleys, in hopes of learning more about the forces that are still creating our planet.

OVERLEAF: *If you are reading this somewhere on the North American Plate, you, the book, and the landscape around you are all creeping along at 1 to 3 cm per year; Earth's plates still move. The Americas slide westward, slabs of Pacific seafloor dive beneath surrounding continents, Europe and Africa converge, and India smashes into Asia. Eastern Africa has begun to rip away from its parent continent, opening a rift valley. Colliding plates raise mountain ranges (blue shading) above and below the sea.*

Old mountain chains (gray), no longer at active plate margins, recall past collisions. Stresses at the edges of a plate or beneath it may create mid-plate rifts.

Slow as it is, plate motion can command our notice—by volcanic eruption and by earthquake. Both take place mostly at plate boundaries. Volcanic eruptions tend to be explosive where plates collide, quieter where they separate. Hotspot volcanoes, built by magma that melts through the plates, still puzzle scientists.

Earthquakes occur on all three kinds of plate boundary: along the slipping faults of translation zones, in deep-diving faults of subduction zones, and in mid-ocean ridges.

In the following pages we will travel four times round the Earth, to see how spreading, hotspots, slipping, and collision shape our planet and our lives.

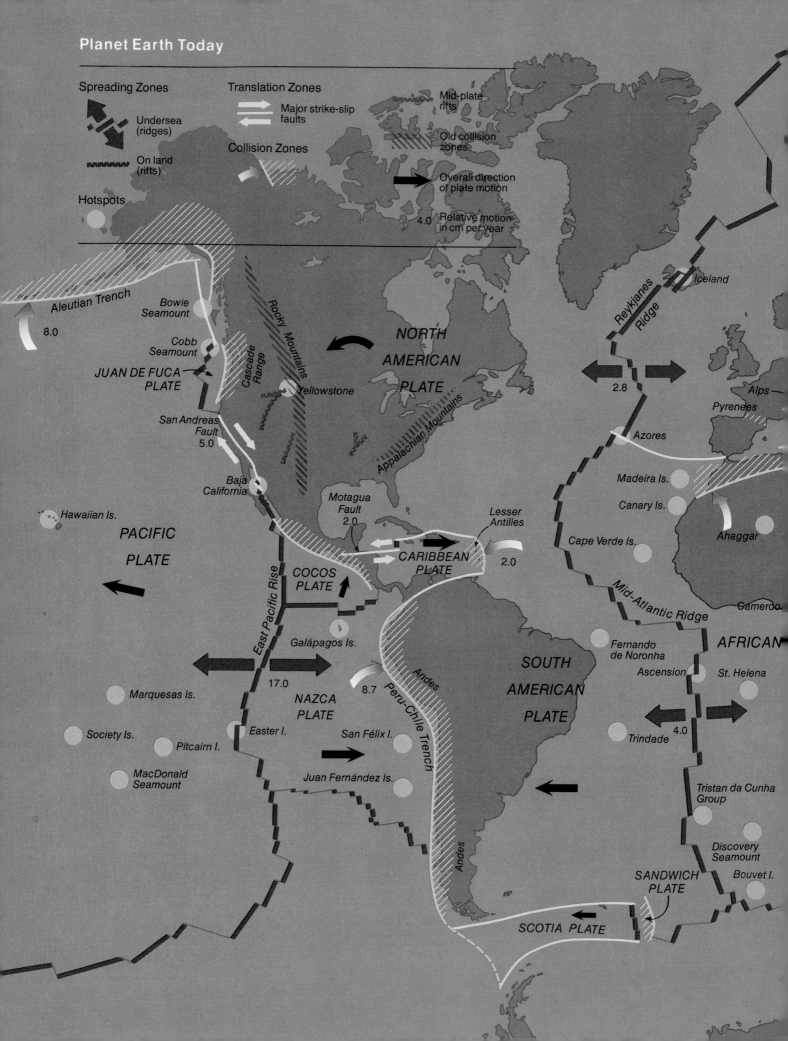

Planet Earth Today

Spreading Zones
Undersea (ridges)
On land (rifts)

Hotspots

Translation Zones
Major strike-slip faults

Collision Zones

Mid-plate rifts
Old collision zones
Overall direction of plate motion
4.0 Relative motion in cm per year

Aleutian Trench
8.0

Bowie Seamount

Cobb Seamount

JUAN DE FUCA PLATE

Rocky Mountains

Cascade Range

Yellowstone

San Andreas Fault
5.0

Baja California

Hawaiian Is.

PACIFIC PLATE

NORTH AMERICAN PLATE

Appalachian Mountains

Motagua Fault
2.0

CARIBBEAN PLATE

Lesser Antilles
2.0

COCOS PLATE

East Pacific Rise

Galápagos Is.

17.0

Marquesas Is.

NAZCA PLATE

8.7

Andes

Peru-Chile Trench

SOUTH AMERICAN PLATE

Society Is.

Pitcairn I.

Easter I.

San Félix I.

MacDonald Seamount

Juan Fernández Is.

Andes

Reykjanes Ridge

Iceland

2.8

Alps

Pyrenees

Azores

Madeira Is.

Canary Is.

Ahaggar

Cape Verde Is.

Mid-Atlantic Ridge

Cameroo

AFRICAN

Fernando de Noronha

Ascension

St. Helena

4.0

Trindade

Tristan da Cunha Group

Discovery Seamount

Bouvet I.

SANDWICH PLATE

SCOTIA PLATE

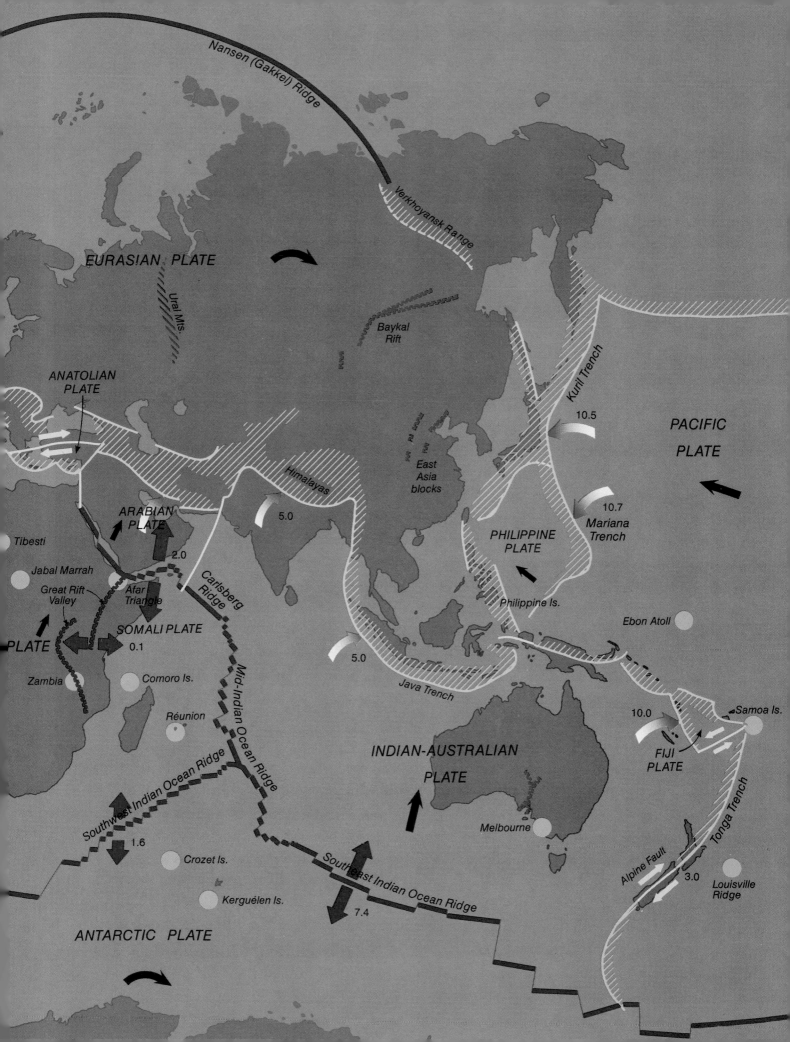

EURASIAN PLATE

Nansen (Gakkel) Ridge

Verkhoyansk Range

Ural Mts.

Baykal Rift

ANATOLIAN PLATE

Himalayas

5.0

East Asia blocks

Kuril Trench

PACIFIC PLATE

10.5

PHILIPPINE PLATE

10.7

Mariana Trench

Tibesti

ARABIAN PLATE

2.0

Jabal Marrah

Great Rift Valley

Afar Triangle

SOMALI PLATE

Carlsberg Ridge

Philippine Is.

Ebon Atoll

PLATE

0.1

Zambia

Comoro Is.

Mid-Indian Ocean Ridge

Java Trench

5.0

10.0

Samoa Is.

Réunion

INDIAN-AUSTRALIAN PLATE

FIJI PLATE

Southwest Indian Ocean Ridge

1.6

Melbourne

Crozet Is.

Southeast Indian Ocean Ridge

Alpine Fault

Tonga Trench

Kerguélen Is.

7.4

3.0

Louisville Ridge

ANTARCTIC PLATE

Spreading

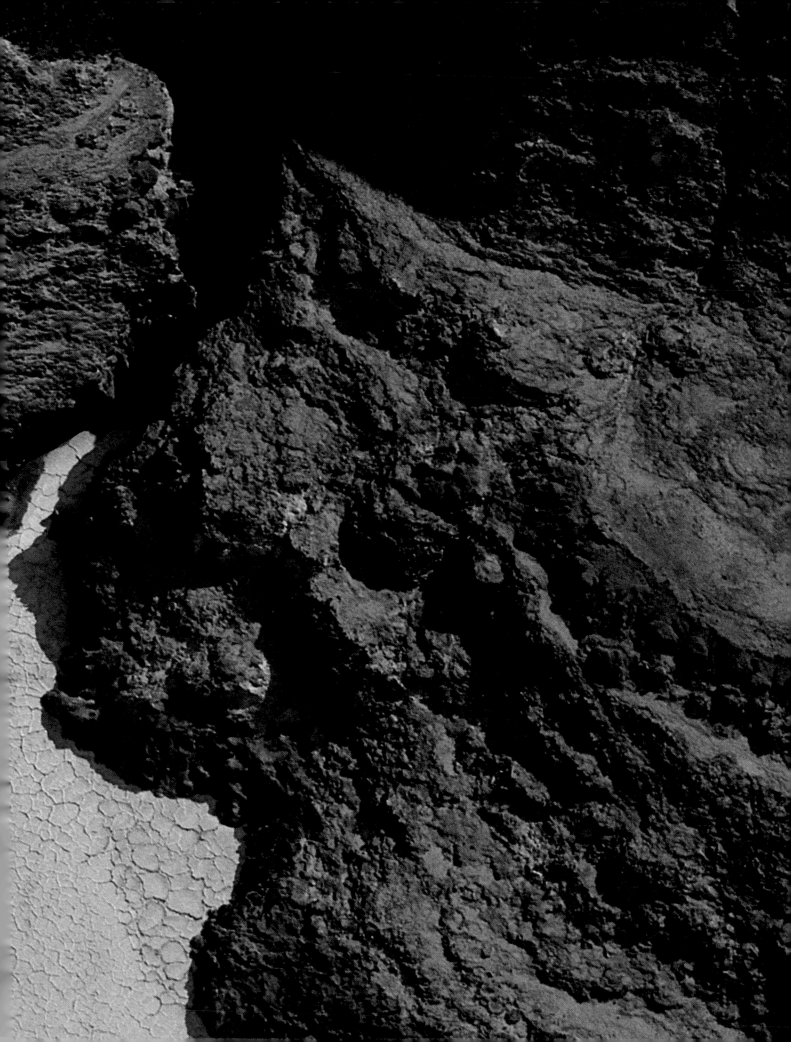

OVERLEAF: *A cleft dozens of feet across breaches hardened lava at Lake Assal in Africa's torrid Afar Triangle. Here widening fissures slowly open the way for a new sea.*

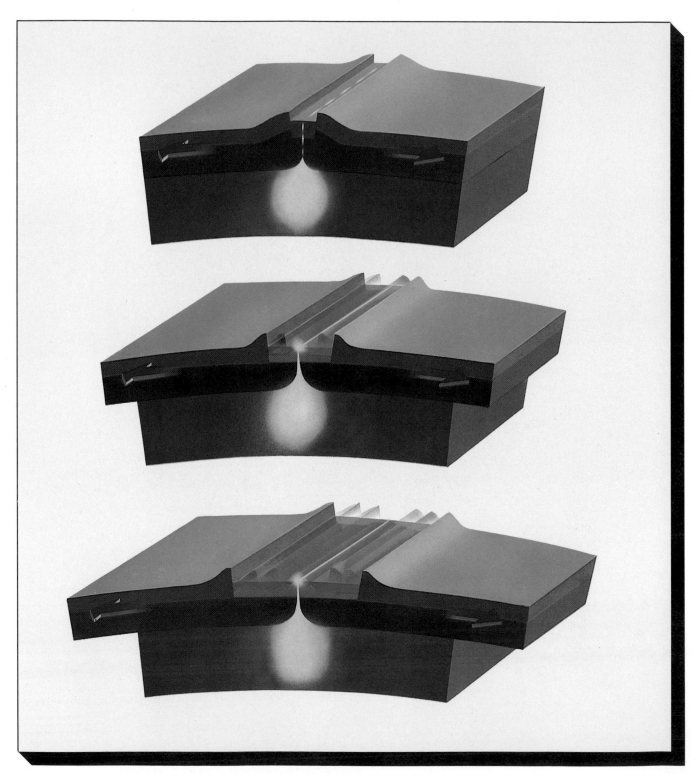

Spreading:
Ocean Maker

Birth of an ocean *begins on dry land when molten rock rises from within the mantle (red layer, opposite). This magma heats the overlying lithosphere (brown layers), which bulges and splits like the crust of baking bread. Along this line of separation, blocks of continental rock subside into a widening chasm, creating a steep-walled* rift valley *(top). Magma oozes from the weakened and fractured valley floor. If spreading continues and the growing rift reaches a coastline, seawater floods in. The upwelling magma builds new oceanic crust, consisting largely of basalt, between the valley walls (center). Eventually the gap between the two continental slabs widens into a full-fledged ocean, thousands of miles across (lower).*

The split has created two plates from one. The boundary is now a mid-ocean ridge, a submarine mountain range buoyed by heat from the mantle. Oceanic crust (light brown) spreads from a central volcanic rift along the ridge axis. As spreading carries away this new rock, it cools and shrinks.

The map (top, right) traces our planet's spreading centers, a network of mid-ocean ridges and rift valleys. Lava squeezing from such fractures slowly adds territory to the plates. Some of the slowest spreading between plates, about 0.1 cm a year, occurs in Africa's Rift Valley; some of the fastest, 17 cm (6.7 in) a year, on the East Pacific Rise off South America.

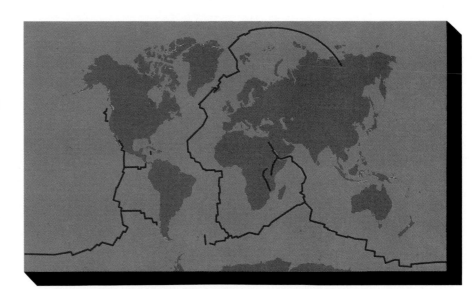

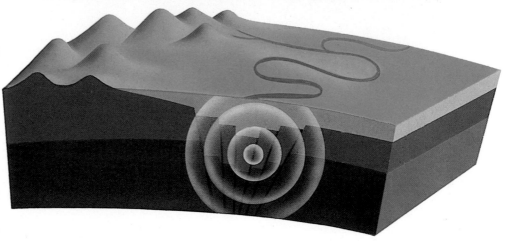

When spreading *gets off to a false start, the result is a "failed rift" (above). Heat bulges and fractures the crust, and blocks of rock drop into the gap—but then the heat abates, and the new rift stops growing. Over millions of years, sediment eroded from distant mountains can completely bury the valley. Out of sight, however, is not out of action; the heavily faulted rift constitutes a zone of weakness, even if far from the edge of a plate. Stresses transmitted* *from the mantle or through the surrounding stable continent can trigger earthquakes along the old fault lines. Three such mid-plate quakes struck in 1811-12 near New Madrid, Missouri, where the Mississippi River flows above a buried rift. Shock waves as strong as those of the 1906 San Francisco quake radiated outward, ringing church bells in Boston and swinging chandeliers in Washington, D. C., drawing rooms.*

shelters Little Magadi, one of the valley's soda lakes. In this semidesert, where Masai cattle graze the scanty grass, whales may someday graze on plankton; geologists expect the Rift to widen slowly into a sea.

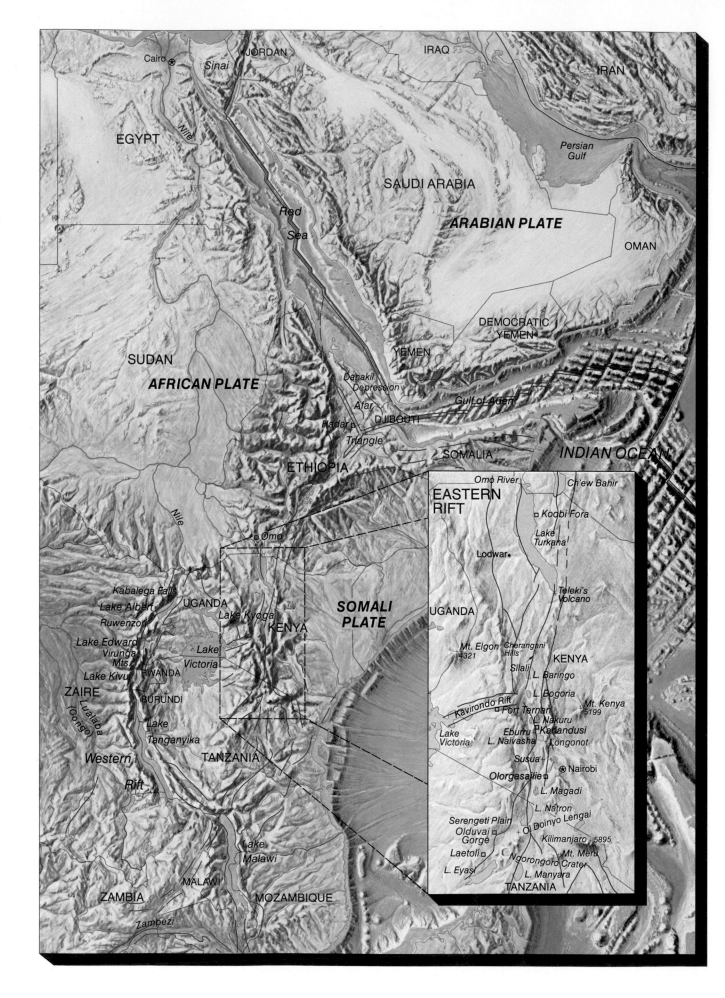

EGYPT

Cairo ⊛

Sinai

JORDAN

IRAQ

IRAN

Nile

Red

Sea

SAUDI ARABIA

ARABIAN PLATE

Persian
Gulf

OMAN

SUDAN

AFRICAN PLATE

Danakil
Depression

Afar

Hadar □

Triangle

DJIBOUTI

DEMOCRATIC
YEMEN

YEMEN

Gulf of Aden

SOMALIA

INDIAN OCEAN

ETHIOPIA

Nile

□ Omo

Kabalega Falls

Lake Albert

Ruwenzori

UGANDA

Lake Kyoga

KENYA

**SOMALI
PLATE**

Lake Edward

Virunga
Mts

Lake Kivu

ZAIRE

RWANDA

BURUNDI

Lake
Victoria

Lualaba
(Congo)

Lake
Tanganyika

Western

Rift

TANZANIA

Lake
Malawi

MALAWI

ZAMBIA

MOZAMBIQUE

Zambezi

EASTERN
RIFT

Omo River

Ch'ew Bahir

□ Koobi Fora

Lake
Turkana

Lodwar •

UGANDA

Teleki's
Volcano

Mt. Elgon
4321

Cherangani
Hills

KENYA

Silali

L. Baringo

L. Bogoria

Kavirondo Rift

□ Fort Ternan

Mt. Kenya
5199

L. Nakuru

Lake
Victoria

Eburru

Kariandusi

L. Naivasha

Longonot

Susua +

⊛ Nairobi

Olorgasailie □

L. Magadi

L. Natron

Serengeti Plain

Ol Doinyo Lengai

Olduvai □
Gorge

Kilimanjaro + 5895

Laetoli •

+ Mt. Meru

Ngorongoro Crater

L. Eyasi

L. Manyara

TANZANIA

Clearly visible from space, the East African Rift System furrows through Africa for 4,500 km (2,800 mi), splitting off a lithospheric slab that scientists tentatively call the Somali Plate. To the north, the Red Sea floods a part of the system where ocean growth has reached the next stage: a wider, deeper valley, paved with seafloor basalt.

Caked soda flats at Kenya's Lake Magadi follow a Rift formula: Add alkaline hot springs to a shallow lake bed and bake in dry heat; replenish springs with a few weeks of rainwater twice yearly. The name comes from the region's Masai language: makat, *soda.*

It is commonly said among the Greeks that "Africa always offers something new." *Pliny*

From the windows of the Cessna, I examined the weird and beautiful patterns of Kenya's Lake Magadi below me. Glaring white soda flats framed an abstract painting in red pastel swirls—algae blooming in the shallow lake water. I was on my way to inspect a mining operation that harvests the thick soda deposits, an unusual and commercially useful by-product of the forces creating Africa's Great Rift Valley, wherein the lake lies.

The long, forking gash of the Rift is where Africa is breaking apart. As the Rift slowly widens, its floor subsides, making way for a future sea. South of Ethiopia, the valley divides into eastern and western branches, as if the crust will not tear cleanly. The Western Rift is lush, with huge, deep lakes and few volcanoes; the Eastern is dry, with smaller, shallow lakes and many volcanoes. (Some people apply the term "Great Rift Valley" to the eastern branch alone.)

North of Ethiopia the valley has already opened and allowed the ocean to enter, forming the Red Sea and the Gulf of Aden, both a part of the larger East African Rift System. Scientists study this system, seeking clues to the birth and growth of oceans. We want to know what Africa can tell us about the opening of the Atlantic 200 million years ago.

A rift valley, or *graben*, forms when the crust separates and a narrow slice of rock slips downward while the sides of the valley bulge upward. In cross section the Rift looks as if the top of an archway had ballooned upward and outward until the keystone dropped partway.

When a graben forms, streams flow into it, pooling in lakes or seeping into faulted rock in the valley floor. What happens to this trapped water depends on climate and topography. Thus one segment of the Rift features lakes saturated with minerals; another, underground streams of heated salt water; a third, rivers captured by freshwater seas. Water in the Rift cascades over waterfalls, vaporizes in steam vents, bubbles in boiling springs, and flows backward through reversed streams. It has formed a deep lake suffused with flammable gas, a shallow lake so salty it's deadly, and a lake where water literally disappears up the chimney. And, I was finding, water throughout this line of valleys gives signs of what the Earth is doing below. Lake Magadi was a case in point.

The Eastern Rift: bitter waters from a startling source

Our small plane descended toward a low ridge that rose from Magadi's shore. Here in the arid Eastern Rift, lakes can evaporate as fast as they fill, and Magadi is an extreme, much of it so dry it can actually be quarried for its soda.

As soon as the pilot, Ian Hughes, set the plane down on a small airstrip, oppressive equatorial heat filled the cabin. We got out and sat in the shadow of the wing to wait for the car from the Magadi Soda Company. And waited. No car. A Masai warrior appeared from the bush with a long spear on his shoulder. Ian talked with him in Swahili and reported that he had come to water

Steps on a Trail Four Million Years Long

The very oldest footprints in the sands of time lead along the Rift Valley—not in sand, really, but in rain-dampened ash sifted from a smoldering Rift volcano. Hominids crossed this area, now Laetoli in Tanzania, about 3.6 million years ago. A team led by anthropologist Mary D. Leakey found the hardened prints in 1978 (right).

These small steps were a giant leap for anthropology, pushing back by half a million years the earliest date then known for hominids that walked upright. In 1982, still older traces—fossil bones about four million years old—turned up in Ethiopia, also in the Rift.

Why does the Great Rift Valley hold so many clues to the roots of

our family tree? For one thing, rifting makes fossils easier to find. Ancient lakeshores hold many fossils, and since lake levels in the Rift rise and fall as the climate changes, shorelines move back and forth, leaving extensive layers of sediment. After long burial there, fossils can reappear as the Rift widens and its floor subsides.

Blocks of crust drop or tilt, and faults slice through the fossil beds. When a block on one side of a fault drops, the higher side erodes, exposing the fossil-loaded layers.

More startling is the possibility that early humans developed in the Rift partly because of the rifting process itself. According to this theory—still conjecture—the uplift

that preceded the opening of the Rift raised forested regions to heights favoring open savanna. This new environment turned tree dwellers into walkers—ancestors of the Laetoli hominids. When the Rift formed, it gave the area a changing mosaic of habitats—hot, cold; high, low; wet, dry—that kept the evolutionary momentum going.

The Masai "mountain of God," —Ol Doinyo Lengai—spews ash that turns into washing soda. The Rift's comparatively slow rate of spreading may permit geochemical processes that account for this unusual ejecta, matched by no other active volcano in the world.

his cattle at a nearby tank. Below the ridge on which we landed, I could see cattle trails leading to it.

I decided to explore. A long trek across a frying-pan African landscape, which turned out to be the company golf course, brought me to the incongruous sight of a British pub outfitted with a shaded swimming pool and an artificial waterfall. The managerial staff was here for the noon break, hence our unnoticed arrival. Beyond the pub was a company town, complete with apartment blocks, schools, hospital, stores, a church, and a mosque.

I retrieved Ian, ate lunch, and then joined Tony Parrish, works manager, in his headquarters office. Tony said that the Magadi Soda Company extracts over 200,000 tons of soda annually from the lake, but added that nature "continues to deposit more than this each year." Centuries of evaporation have left a deposit of soda, known as *trona*, estimated to be 200 meters (670 ft) thick. Kenya exports the processed soda mainly to other Indian Ocean nations, where it serves various industrial purposes from soap and detergent manufacture to glassmaking and water softening.

Tony drove us past the processing plant and out onto the solid soda crust. Sunlight bounced off the flat white surface, intensifying the heat. Huge dredging machines were cutting wide channels about three meters (10 ft) deep into the trona. In slurry form, harvested trona was pumped to the processing plant. As the dredges progressed across the flats, soda-saturated water that Tony called "liquor" quickly filled the excavated channel. The company hopes to find a way to extract

Young Masai women gather to celebrate a birth. While Africa modernizes, the haughty Masai cling to customs they have followed throughout the four centuries the tribe has lived in the southern part of the Eastern Rift. Tribal lore reflects Masailand's volcanic geography; generations of storytellers have told of how their people ascended from a barren crater into green lands above. Masai oral history dates other legendary events according to whether they happened before or after this milestone Ascent.

soda from the liquor, which holds fewer impurities than solid soda.

Where does all this soda come from? The answer lies in the Rift to the south, on the slopes of the volcano Ol Doinyo Lengai, the Masai's "mountain of God."

A peak of a different color

Ol Doinyo Lengai has thrown out clouds of brown ash several times this century, but after each eruption the mountain mysteriously turned white. In 1960, when an eruption also produced a lava flow, the Geological Survey of Tanganyika sampled it and discovered the volcano had disgorged a previously unknown rock made largely of an alkaline salt, sodium carbonate.

Analyses of lava and ash explain why the color changes. After a rain, the carbonates in the fresh ash dissolve, recrystallize, and shroud the mountain in a white crust—of ordinary, old-fashioned washing soda. Eventually more rain cleans the soda away, leaving only insoluble ash, and Lengai returns to its normal gray-brown hue. It's the only active volcano known to erupt such concentrations of soda.

Earth scientists are still trying to understand why the Rift produces this phenomenon. In experiments with clay models, geologists have discovered that the surface width of a graben is often about the same as the thickness of the crust in which it occurs. The Rift Valley varies from 30 to 50 kilometers in width, and the continental crust through which it runs seems to be equally thick. That means magma from the mantle rises through a lot of crust to reach the surface, and the deeper the source of the magma, the greater

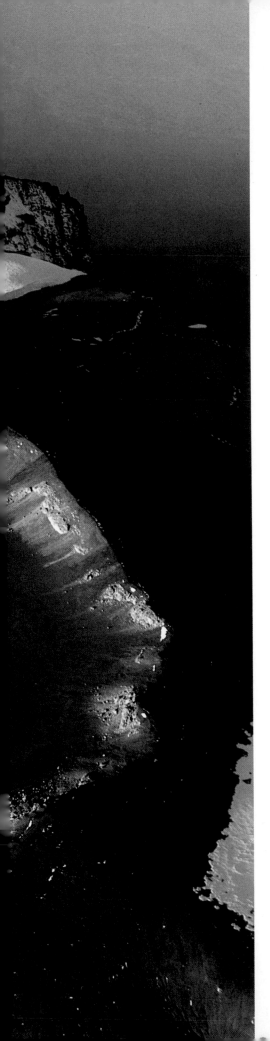

The thin, arctic breath of high altitude chills the top of Africa: Kilimanjaro's icy summit and the inactive "ash pit" in its center. At 5,895 m (19,340 ft), the volcano is Africa's highest mountain. Kilimanjaro and several other lonely giants soar next to the Rift, all products of the same deep-seated geological forces.

the pressure where it melted. (In thinner seafloor crust, by contrast, the magma wells up from lesser depths under lower pressure.)

High-pressure melting produces alkaline magmas. Compared with other continental volcanoes, magma deep under Lengai holds unusually high traces of carbon dioxide, which makes carbonates. As this magma rises, dropping pressure and temperature allow it to separate into its constituent parts, like oil from vinegar. The lighter, soda-rich carbonates rise to the top and eventually erupt on the surface.

But why does this happen in the Rift? For one thing, carbonate magma does not appear where plates separate quickly. The Rift Valley, however, opens very slowly, no more than a millimeter a year. That's less than 5 percent of the rate at which the Atlantic widens. Also, mantle rock tends to hold the necessary amounts of carbon dioxide only if the rock has been undisturbed for millions of years. Since Africa has moved little in recent ages, no passing plate boundaries have disrupted the mantle below.

Magadi and its neighbor to the south, Lake Natron, owe their huge soda deposits to Ol Doinyo Lengai, but all the Rift volcanoes are unusually alkaline, because their lavas formed deep, at high pressure. Several million years of such volcanism have laced the rock and soil with alkaline minerals, and this is why many other lakes of the Eastern Rift are also soda filled. Runoff in the rainy season leaches out soda compounds and seeps into fractures that riddle the subsiding valley floor. There the water often picks up heat from subsurface magma, dissolves still more soda from the rock, and reemerges in hot springs—which empty into the soda lakes.

Springs with many uses

In fact, hot springs pepper much of the valley floor. At Olkaria, in the Rift north of Lake Magadi, a new power plant is using this geothermal activity by tapping an underground reservoir of hot water under high pressure. My tour of the drilling operations required ear protectors against the thunder of escaping natural steam. The plant is expected to generate 45 megawatts, but local residents were already putting it to work by washing their clothes in the hot runoff from the steam wells.

In contrast to the sophisticated Olkaria plant, a farming village on a nearby volcanic hill called Eburru relies on what may be the world's simplest geothermal technology—needing only some cement, an oil drum, and some downspout piping.

The Kenyan government wanted to settle people at Eburru to take advantage of the fertile volcanic soil, but there was no reliable source of water. There were, however, numerous *fumaroles* releasing water vapor from deep inside the Rift. The problem: How to trap the vapor using only equipment on hand in a poor country. Solution: Cement an oil drum over the vent, cut a hole in its side, and run a few dozen feet of galvanized pipe uphill from the hole. The hot steam enters the drum and begins to escape out the piping. But as it drifts along, it condenses on the cool sides of the pipe. This distilled water drips back via a second pipe into a holding tank, whence residents can draw it off for use. *(continued on page 80)*

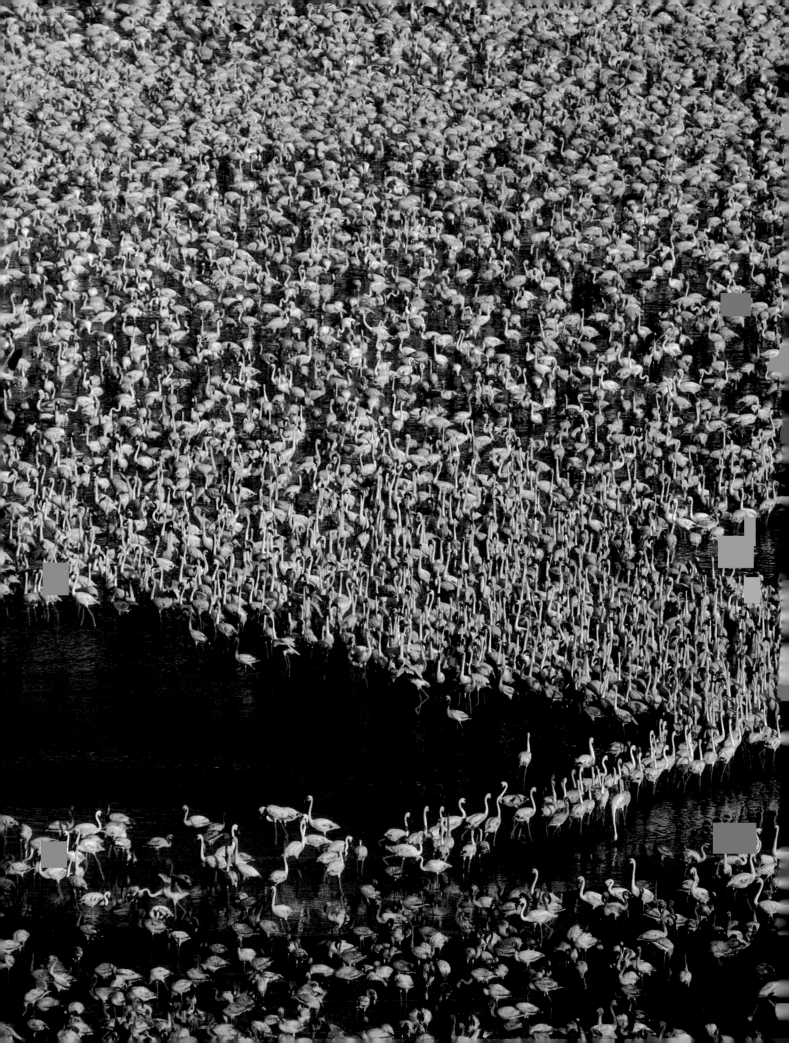

Of Feathers, Soda, and Algae

Massed flamingos owe their pink of health to the soda lakes of the Eastern Rift, and to the algae that grow there. A model of ecological simplicity, the soda lakes nourish fast-growing algae that feed the flamingos; flamingo droppings in the lakes promote algae growth. Flamingo feathers get their hue from pigment in the algae.

75

OVERLEAF: *Close-up beauty of natural soda earns Magadi its nickname, "the crystal lake." This material, called* trona, *knits itself into a solid crust strong enough to walk on.*

2ND OVERLEAF: *A survey plane skims the eerie immensity of Tanzania's Lake Natron during a full algae bloom. Algae create this crimson tide after rainfall coats dense, soda-laced layers with a film of fresh water. Gas and liquid well up in swirls of caustic lye. Shallows in mid-lake provide the main breeding ground for the Rift's flamingos, a secret that Natron's vast, corrosive soda flats kept hidden until aerial reconnaissance in the 1950s found the nests. The Masai, unable to see the nesting area through miles of shimmering mirage, assumed young flamingos sprang from the lake full-grown.*

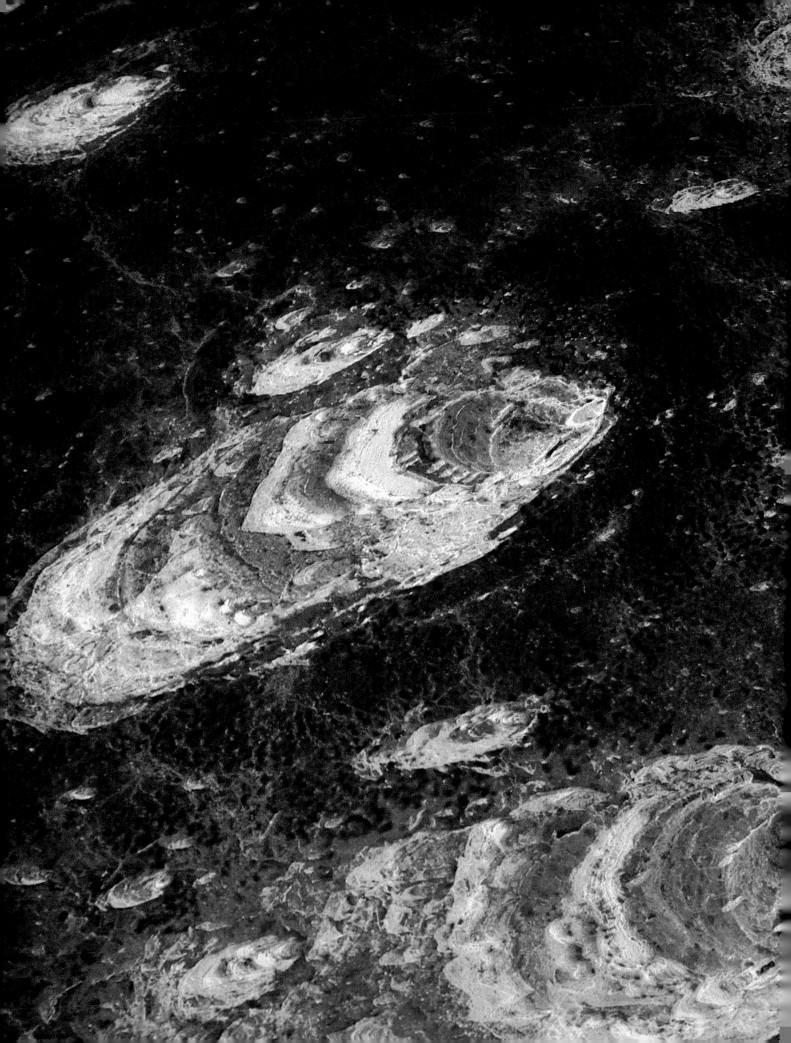

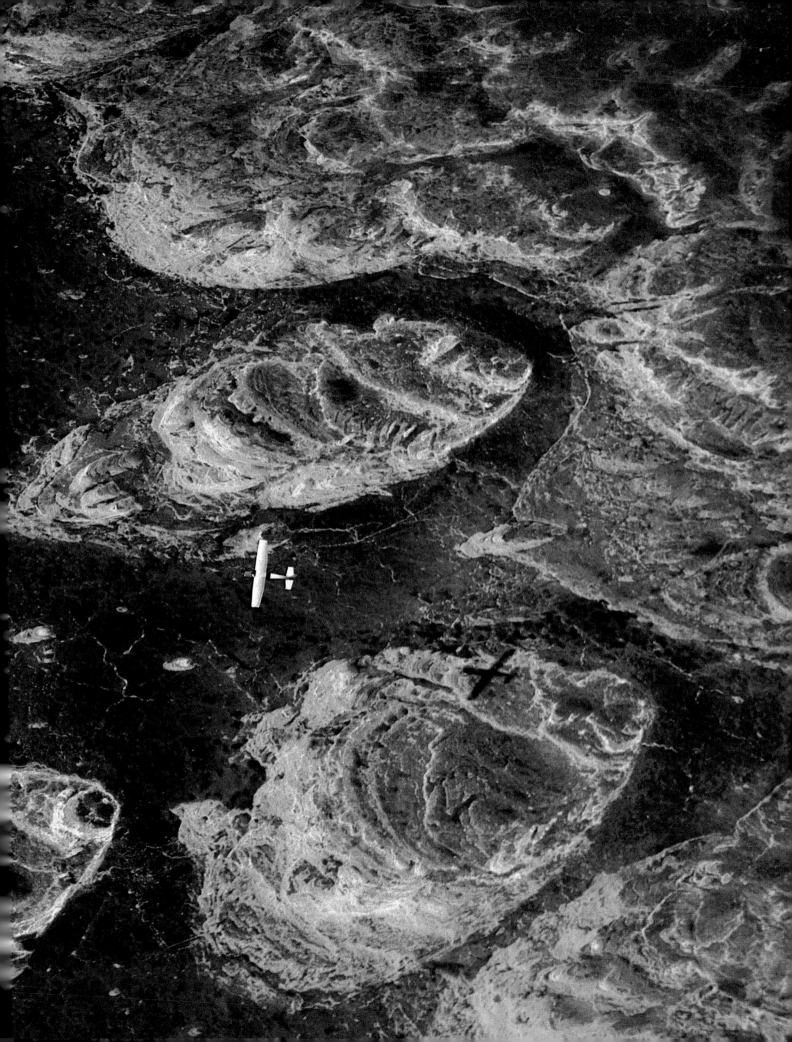

Friend turned foe, Magadi soda entraps a flamingo chick. Usually, fetid soda-lake waters repel most predators and competitors, as in hot-spring-fed Lake Bogoria (below). But if the concentration of soda increases, a crusty deposit can form on the birds' legs. These deadly anklets immobilize the chicks, which starve.

A realm of bats and baboons under the lava

Eburru is one of several inactive volcanoes that rise from the floor of the Rift northwest of Nairobi. Igor Loupekine, a Nairobi-based geologist, took me to another, an overgrown crater called Susua. When Susua last erupted, hundreds or thousands of years ago, rivers of thin lava pouring down its slopes formed tubes, hardening on the outside while the still-liquid interior drained away. Hollow lava tunnels like these can go on for miles. In the 1950s, Mau Mau insurrectionists used the tunnels to hide from the British.

Igor and I prowled around in the grass and bushes near the rim of the crater until he found an opening where the roof of a tunnel had fallen in. (The bones of an unlucky rhinoceros are said to lie at the bottom of one of these holes.) A climb down a tumbled rock wall put us in a tube that must have been at least 8 meters (26 ft) around, the largest I had ever seen. After twisting underground for a couple of turns, it opened into a cathedral-like cavern, lit by bright daylight streaming through a couple of holes in the roof. A stale, sour stench assailed our noses.

Igor pointed to a pyramid of heaped rock, fallen from one of the holes. "Baboons sometimes come here," he said. "They like to slide down the rocks." The stones of the rock pile were smooth, as if worn by centuries of hard use. The rank odor rose from the accumulation of bat droppings on the cavern floor.

We returned to clean air on the surface, and I looked out over the valley. Susua and the other volca-noes within the Rift have erupted great volumes of lava into a graben that is not widening very quickly, so volcanic debris has filled the valley nearly as fast as its floor subsides. By one estimate, the Eastern Rift holds 700,000 cubic kilometers of volcanic rock—enough for a belt of lava across the United States 50 miles wide and a mile deep.

From Tanzania in the south to Ethiopia in the north, lakes of varying size spot the floor of the Eastern Rift. All are shallow—big puddles atop the thick volcanic fill.

With a geochemist colleague, Harmon Craig, I spent a couple of days exploring the area around Lake Baringo. It lies in a section of the valley known as the Gregory Rift, named in honor of John Walter Gregory, the British geologist who proved that this great rent was the product of crustal collapse. Scientists who came after him also identified a small branch graben, the Kavirondo Rift, which extends to Lake Victoria. Such branches seem typical of rift systems; when the main rift opens into a sea, these smaller "failed rifts" remain etched in the rock of the new coastlines.

Baringo, having no significant in-flow from hot springs, is a freshwater lake. Just south of it lies a soda lake, Bogoria, whose shorelines abound in alkaline hot springs—and in flamingos, which congregate in soda lakes to eat the algae there.

Harmon was sampling geothermal springs throughout the Rift, testing for gases which, he says, "come straight from the mantle" and reveal chemistry similar to that found in oceanic spreading centers.

We went to the base of the volcano Silali, where some of the springs

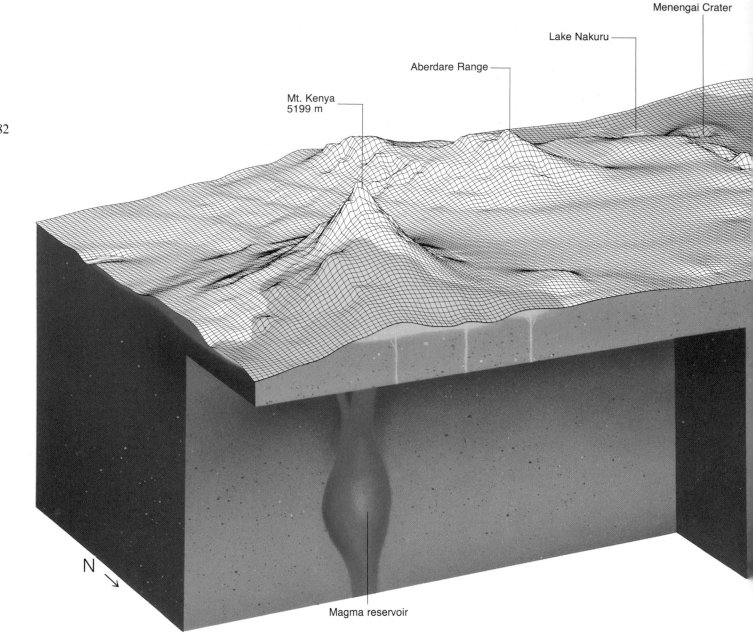

Mt. Kenya
5199 m

Aberdare Range

Lake Nakuru

Menengai Crater

N

Magma reservoir

John Gregory's Rift

Computer-rendered topography adds texture to a cross section of the Eastern Rift in mid-Kenya. Europeans first set eyes on this great valley less than a hundred years ago. One of them, a British geologist named John Walter Gregory, discovered the nature of the Rift by making a laborious traverse near Lake Baringo in 1893. Gregory sampled rock first from high on the eastern valley wall (left side), then from the much lower "Kamasia block," jutting from the valley floor. The strata matched. He concluded, "these valleys were not formed by removal grain by grain, by rivers or wind, of the rocks which originally occupied them, but by the rock sinking in mass, while the adjacent land remained. . . ." The

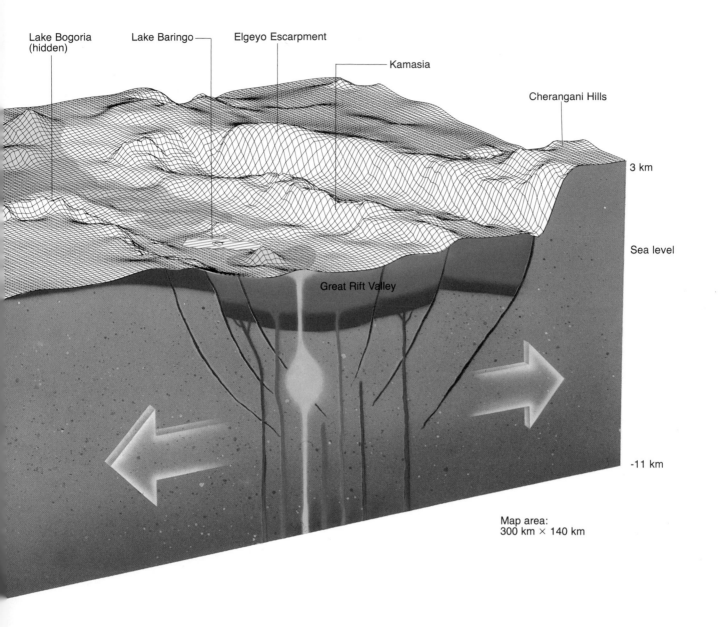

Lake Bogoria
(hidden)

Lake Baringo

Elgeyo Escarpment

Kamasia

Cherangani Hills

3 km

Sea level

Great Rift Valley

-11 km

Map area:
300 km × 140 km

Great Rift Valley was a true graben, *a feature created where a slice of Earth's crust falls as crust on each side pulls away. From the Cherangani Hills, the drop is almost 3 km (2 mi).*

The weak, extended crust allows magma to rise. Volcanoes inside the "Gregory Rift" regularly re-layer the valley floor in lava. One of them made the crater Menengai,

Masai for "place of the corpses." Here in the 1800s, one Masai army vanquished another by forcing the enemy warriors over the rim. A far larger volcano, Mount Kenya (now extinct and eroded) grew, like Kilimanjaro to the south, to an exorbitant height outside the Rift.

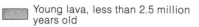 Young lava, less than 2.5 million years old

 Lava 2.5 to 12 million years old

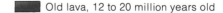 Old lava, 12 to 20 million years old

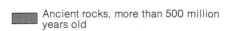 Ancient rocks, more than 500 million years old

84

he sought flowed together to form a hot waterfall as tall as a tree. It was the end of the dry season in the Rift, and the equatorial sun glared on a tired, scraggly landscape. The falls looked cool and inviting—until we stepped into the river. Harmon's thermometer found the water to be a barely tolerable 45°C (113°F)!

The Jade Sea
Because of climate fluctuations and shifts in the uneasy crust below, lake levels in the Eastern Rift rise and fall. The current semiarid climate has shrunk Lake Turkana along with the others, but it remains the biggest Eastern Rift lake by far. It has no outlet and its water is brackish and bitter. Still, it supports a wide variety of fish and birds, as well as some 22,000 crocodiles, the densest concentration in Africa. Much of the lake's aquatic fauna resembles that of the Nile, evidence that an outlet to that river once existed. Abundant green algae inspired the lake's nickname, the Jade Sea. What saves it from becoming a soda flat like Magadi is the Omo River, which injects fresh water from the Ethiopian highlands.

The lake's ups and downs have made the region an ideal storehouse of fossils. Ancient lake sediments to the east now nourish a dusty scrubland called Koobi Fora. Here anthropologist Richard Leakey discovered in 1972 a hominid skull two million years old. In the days when the owner of "skull 1470" walked here, the scrubland was probably a forest on the shore of an enlarged, sparkling lake, its waters kept fresh by the outlet to the Nile.

Despite its size and the length of human residency here, Lake Tur-

kana went undiscovered by Europeans until 1888, when Hungarian Count Samuel Teleki arrived on a hunting expedition. Teleki named his find Lake Rudolf, for the crown prince of Austria. With independence, the Kenyans renamed it Turkana, after the region's major tribe.

Ian Hughes and I explored the area, stopping to refuel his plane in Lodwar, the dusty main town of the Turkana region. The place smacked of an Old West frontier town, complete with primitive saloon. Refueling required rolling out a drum of gasoline and pumping it by hand through a strainer into the tanks.

We took off and headed for the southern end of the lake. Rift volcanism had reshaped this area some 10,000 years ago, when a volcanic barrier arose and chopped off the lake's southern tip.

We flew over this natural dam, some of it clad in fresh-looking black lava. During Count Teleki's trek here, his aide wrote of "an overwhelming smell of sulphur and chlorine, which soon made our men cough. . . ." Below us stood a cone the Europeans named Teleki's Volcano; the Turkana, though, are said to call this tract Lugugugut, "the place that is burnt." On the other side we found a drying soda lake, all that remained of Lake Turkana's southern reaches. Far below, formations of pink dots skimmed the shallow, bitter water—flamingos.

North of Lake Turkana the Rift squeezes through the Ethiopian Plateau, opening on the far side into a searing wasteland called the Afar Triangle. Here the threat of an ocean invasion is no vague geological prospect millions of years in the future; here it is under way.

OVERLEAF: *Ngorongoro wildebeest parade against the hazy backdrop of 610-m (2,000-ft) crater walls. Wildebeest cross these ramparts on yearly migrations to and from the Serengeti Plain. Even though outside the Rift, the Serengeti's lush grasslands—foundation of its ecosystem—thrive in soil enriched by ash from Rift volcanoes.*

Teamwork on the floor of hell: Afar tribesmen may endure temperatures higher than 49°C (120°F) as they pry salt from dry lake beds in Ethiopia's Danakil Depression, 120 m below sea level. The area occupies the northern reaches of a geological wonderland named for these people—the Afar Triangle.

The Afar Triangle, Africa's volcanic floodgate

Our Land Cruiser jounced along a track through the western corner of the tiny, arid nation of Djibouti. In the rocky, parched countryside nomads watched their goats, and camels roamed unattended, looking for something to eat. Platoons of dust devils marched across the distant flats. The region had had no significant rainfall in four years. This, a local chief had told us, was a drought; usually it rained every two years. Down the road, we passed a camel carcass, mummified.

Djibouti seemed to me what all Earth might have looked like in its stark youth, and I was glad to have a guide, my friend and colleague Jean-Louis Cheminée of the French National Center for Scientific Research. Our goal was Lake Abbe, on the Ethiopian frontier.

Djibouti lies entirely within the Afar Triangle, a complex tangle of faults, rifts, and volcanoes. Here the spreading zones that bound the African, Somali, and Arabian Plates join in what geologists call a triple junction. Two avenues of the intersection have already flooded: the Red Sea and Gulf of Aden. The third, the Rift Valley, is just beginning to fill. Harmon Craig's research here leads him to believe that the Afar is not only a spreading center, but a hotspot as well, supplied by very deep-seated magmas. Many other scientists agree, and wonder whether hotspots actually begin the rifting process.

Lake Abbe interested us because of the hollow, chimneylike spires of travertine that rise from its flat

The Afar Triangle began to form when Arabia undocked from Africa (top), opening two rift valleys and pulling a sliver of continental crust free in the process. Ocean filled the growing gaps, but continuing earth movement and volcanism dammed off the low-lying Afar desert area. It has not been desert long; at the foot of the Ethiopian escarpment one expedition found a 200,000-year-old stone ax crusted with seashells from the earlier flooding. Today the Afar Triangle lives on borrowed time; the striped zone, geologically more like oceanic than continental crust, will again succumb to the sea when rifting eventually disrupts high ground to the east.

shores. The spires, many higher than a house, sit atop hot springs rich in bicarbonate. Steam from below quickly evaporates in the searing air, leaving a deposit of limestonelike travertine.

"Lake Abbe is fed by the River Awash, whose source is on the Ethiopian highlands," Jean-Louis said. "The lake is very alkaline and very shallow, about two to four meters only." Lake Abbe is in fact the final resting place of the Awash, which never reaches the sea. The river flows instead into a network of lakes of which Abbe is the last, and its waters either evaporate or sink into faults, to reemerge in hot springs or from the chimneys.

Jean-Louis said the chimney vapor was visible only in the early morning (it evaporated too fast in the heat of day), so we camped for the night and rose just after dawn to see wisps rising from the spires— the Awash going up in steam.

Enter the sea

Less than 90 kilometers from Lake Abbe is the place where the Great Rift Valley meets the ocean: a small graben called the Assal Rift. Here a new sea has actually begun to form.

The Assal Rift is a dry, lava-floored trough between the Gulf of Tadjoura and super-salty Lake Assal, 156 meters (512 ft) below sea level to the west. Rising high enough to block the ocean from the lake, the 12-kilometer stretch of exposed seafloor constitutes a kind of geological twilight zone, caught in time between land and ocean.

Jean-Louis and I took in the view from the rim of the valley. Down the valley's center runs a line of small volcanoes draped with fresh lava—not the frothy soda ash of Ol Doinyo Lengai, but basalt, heavy and black, the stuff of seafloors. This is the very boundary along which the plates separate—one of the most active zones in the Great Rift Valley. The most recent flow, said Jean-Louis, erupted from a series of fissures in November 1978.

We guided our Land Cruiser down into the valley, following a track made of lava chunks the size of cinder blocks, and found a campsite with a distant view of Lake Assal. As we set up our tents, I asked Jean-Louis about the small spiral shells, bleached white, that lay everywhere in the dust and rocks around us. How did they get here, 115 meters (377 ft) above sea level? He said they were from the lake, which once stood much higher than it does now. The lake level fell, and a temporary rise in the valley floor had lifted the shells to their present unlikely altitude. They have lain here in the desert for 8,000 years.

The Assal Rift offered us a fantastic opportunity to examine seafloor spreading unencumbered by the ocean. We drove to the Gulf of Tadjoura to investigate the seaward end of the rift. Under a blazing sun, we hiked across a raw, tumbled lava field to the edge of a deep fissure. At the bottom of the cool humid crack seawater flowed, heading toward Lake Assal.

We drove along the route this water takes as it seeps underground through the rift. At one point we stopped to inspect a series of steam vents. The steam was from boiling salt water; some four kilometers beneath us, seawater percolating through the porous crust passed near a hot magma chamber. The

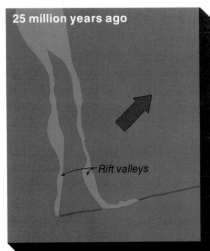

25 million years ago

Rift valleys

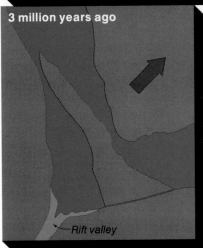

3 million years ago

Rift valley

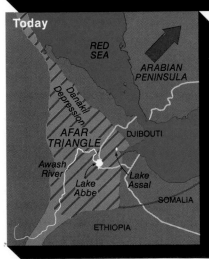

Today

RED SEA

ARABIAN PENINSULA

Danakil Depression

AFAR TRIANGLE

DJIBOUTI

Awash River

Lake Abbe

Lake Assal

SOMALIA

ETHIOPIA

In the Crucible of Creation

Raw seafloor cooks under a desert sun in Djibouti's volcanic Assal Rift. Here Africa comes unzipped. From above the Gulf of Aden coastline, this view looks inland along the length of the graben— the ocean's line of attack. In the background, at the other end of the trough, salt rings landlocked Lake Assal, 156 m (512 ft) below sea level. Ocean water seeps through the fractured lava under the valley and feeds the lake; dry, hot air evaporates it, leaving the salt.

OVERLEAF: *Pools of emerald and jade—brine seeping from fissures —lie clasped in settings of oxidized rock in the Danakil Depression. Iron chlorides suffuse the pools, whose green hue grows deeper as the water evaporates; rust tints the surrounding deposits, all photographed here from an altitude of 60 m (200 ft).*

2ND OVERLEAF: *Travertine chimneys, molded by bicarbonate springs, sprout from Lake Abbe's broad shoreline. Their linear patterns reveal underground faults from which mineral-bearing vapor rises. As the vapor evaporates, it deposits a spire of minerals around each vent. Djibouti portrays the formations on its 500-franc note.*

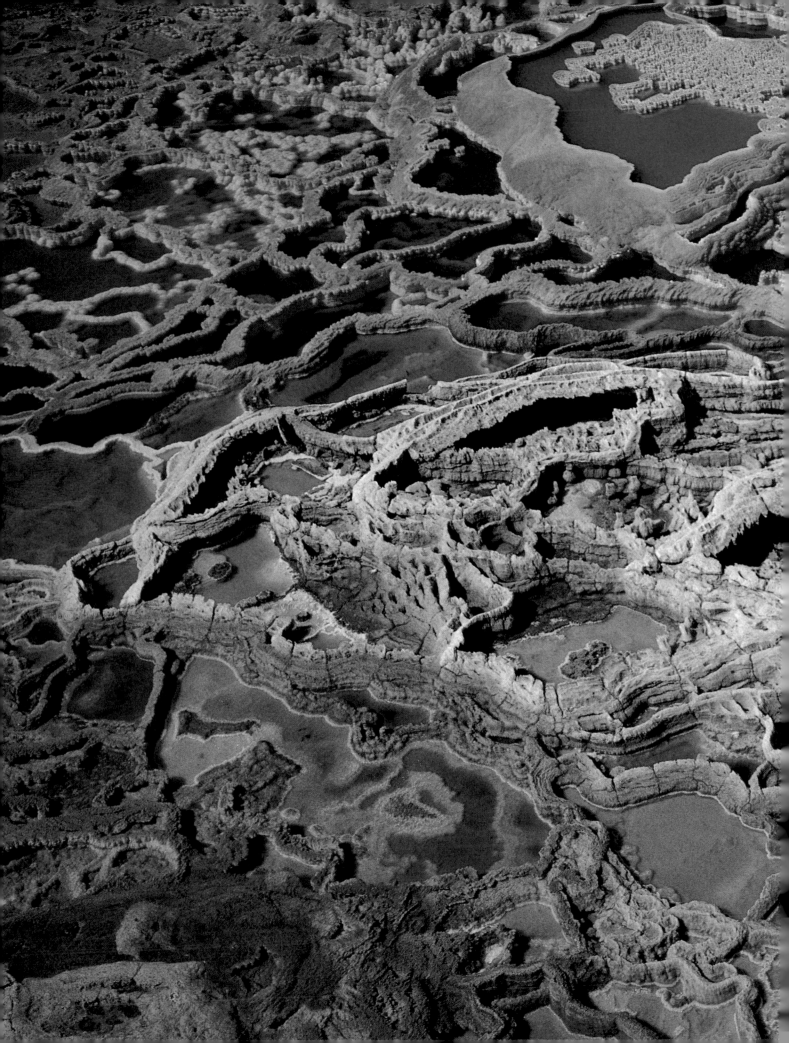

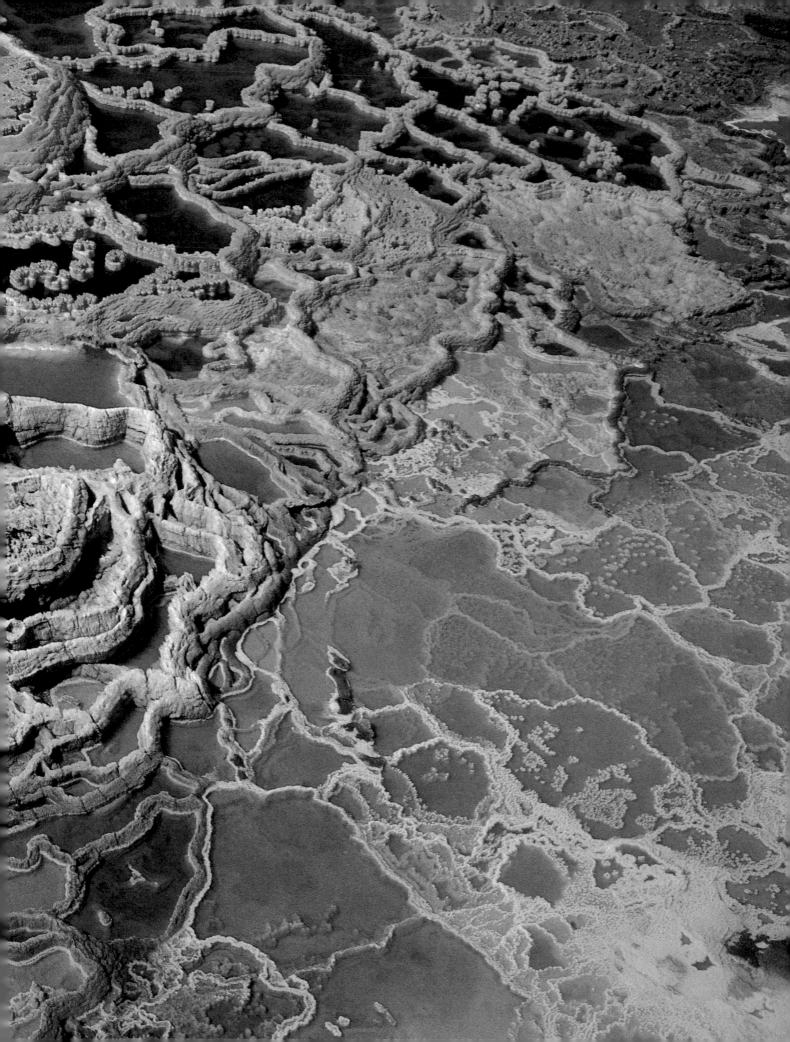

water turned to steam and issued from these surface vents. It condensed on our clothes, providing temporary relief from the dry heat.

A long rocking Land Cruiser ride took us down to the inland end of the rift, and Lake Assal. Along the shore, seawater issued from cracks in the lava, having completed its trip beneath the rift floor. Some of these springs were cool, but others arrived steaming hot from a close pass by the magma chamber.

Were it not for the desert climate, Lake Assal would soon fill the depression in which it lies, as it has in wetter times long ago. Now, however, evaporation is so intense that the incoming ocean water is sucked up into the dry air, leaving a thick crescent of salt around the lake. Most of Assal is so saline that any life washed into it quickly dies. But in the small coves where seawater flows in, we did find living things: the same snails whose ancestors' shells lay around our campsite.

The Western Rift, valley of deep waters

In Tanzania, partway between the Afar Triangle in the north and Mozambique in the south, the Great Rift Valley encounters the Tanganyika Craton, one of the oldest rock massifs of the continent. Unable to penetrate this ancient shield, the Rift behaves like wood splitting around a knot. The Eastern Rift peters out at the craton, but the Western Rift detours around it, cutting a green crescent 3,400 kilometers (2,100 mi) long, filled with narrow, deep lakes.

This watery valley forms a natural boundary for nine nations, from

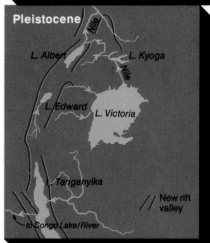

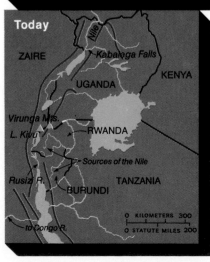

Sudan to Mozambique. Unlike the Eastern Rift, the climate in most of the Western is wet. Water draining into the grabens doesn't evaporate as fast as rain replaces it, so lakes keep growing until they find an outlet over the side of the trough that holds them. The Western Rift cradles many of what are called the Great Lakes of Africa, and they deserve the name. Scientists working from a boat on Lake Malawi in 1981 complained of 12-foot seas. Lake Tanganyika is the world's longest lake and the second deepest.

Once again, water tells the geological story. These huge lakes reveal the dominant force at work in the Western Rift: movement of the land itself. Volcanism, though present, yields in importance to tectonics. Tectonics made the lakes.

When the Rift began to take shape about 20 million years ago, in the Miocene epoch, entire regions of the continent swelled into huge, gentle domes that split across the top like overripe tomatoes, forming the rift valleys. Ethiopia is one such dome, the highlands of Kenya another, and the area centered north of Lake Tanganyika another.

Before doming began, most streams here drained westward into a gigantic lake that filled the Zaïre (Congo) River Basin. But when the Western Rift opened, it captured this part of the Congo watershed— stream piracy, geologists say. For a while the former tributaries of the Congo Lake emptied instead into the new graben, forming lakes. But as uplift continued, the tributaries began to flow backward, following their own streambeds east instead of west. These reversed rivers still flow that way today. Caught

Giant natural spillway, Kabalega (Murchison) Falls thunders over the Rift escarpment in Uganda. Uplift to the east dammed Nile headwaters until they rose high enough to find a new outlet—this 40-m (130-ft) drop on the Victoria Nile. Europeans named the falls for geologist Roderick Murchison, historian of Paleozoic fossils.

100 between the slopes of eastern and western domes, they flooded a former marsh and there created Africa's largest lake, Victoria.

The lake that brews beer

The deeply eroded highlands of the western dome now support the small, densely populated countries of Rwanda and Burundi. The road west from Kigali, the tiny capital of Rwanda, took me through a countryside of villages and farming plots cut into lush, steep hillsides. People were everywhere. Rwanda, called "land of a thousand hills," owes its fertility—and ballooning population—in part to the Rift's volcanic soil, now sadly overworked.

A curve in the road suddenly produced a spectacular vista: the Virunga Mountains. These volcanoes—the only ones in the Western Rift—tower as high as 4,510 meters (14,787 ft). Six of the eight peaks are now silent and jungle clad. They are famed as the last refuge of the rare mountain gorilla.

Two peaks, over the border in Zaïre, are still active—Nyamlagira and Nyiragongo. Their lava flows dammed an ancient, north-flowing river to create Lake Kivu, the only major Western Rift lake shaped by volcanic, not tectonic, action.

Lake Kivu contains a large amount of dissolved methane gas. The concentration increases with depth. The methane may come from organic material trapped in the lake, or indirectly from volcanic gases. Lake Kivu stores so much gas that from 1963 to 1976 European settlers extracted it to fuel a brewery on the shore.

I visited the old extraction plant, which the Rwandans were rebuilding. A large pipe ran into the lake, and down into the methane-charged depths. A suction pump would bring up the gas for processing. Its final destination? The same brewery the Europeans used.

Kivu may hold the strangest fuel, but scientists now see signs of conventional reserves elsewhere in the Western Rift. Expeditions probing Lake Tanganyika have found that much of the Zaïrian (western) shore is leaking hydrocarbons—crude oil.

Rifts present and past

Is the Great Rift Valley really a window on the past, a replay of Pangea's breakup some 200 million years ago? After all, the Rift is spreading much slower than the pre-Atlantic rifts of Pangea.

Yet in Scotland the eroded cores of giant volcanoes stand, volcanoes that rose not far from the infant Atlantic, as Kilimanjaro rises outside the Rift today. We have found branch rifts like the Kavirondo in Atlantic seaboards; the Connecticut River flows in one. And buried in the continental shelf off New England lie salt deposits. Relics of a giant Assal-like lake? We think so.

And how will Africa break apart? Geologists theorize that in 50 million years or so the widening Rift will have whittled off a long slice from Somalia south to Mozambique, possibly including South Africa as well. Where the fresh waters of Lakes Tanganyika and Kivu now gleam, an arm of the ocean will enter. And beneath its waves, dense basaltic lava will slowly build new seafloor from a central ridge.

To see that process today, geologists have to go to the ocean floor—and that's exactly what we've done.

The Mountains of the Sea

Chariot of discovery, the submersible Alvin *takes scientists to realms of Earth never before seen, probing our world's longest mountain range, the Mid-Ocean Ridge. Here* Alvin, *hovering 2.5 km down in the Pacific, explores a pillow-lava field at the Galápagos Rift.*

Steam from hot mineral springs wreathes the steep streets of Furnas, a health spa on São Miguel in the Azores. Volcanoes rising 4 km from the seafloor created the nine islands and gave them their fertile soil. The research fleet of Project FAMOUS sailed from this island to explore the Mid-Atlantic Ridge in 1973 and 1974.

A brimstone bakery beside a crater lake cooks dinner for villagers on São Miguel. Tectonic activity opens escape routes for magma from deep in the Earth; its heat seeps into the soil and cooks the cloth-wrapped pots of pork and cabbage or chicken and vegetables in three hours. Sulfur fumes add piquancy to the food.

Odin flung the serpent Midgard, Loki's second son, into the deep sea which surrounds the whole world, and it grew so large that it now lies in the middle of the ocean round the earth. . . .
From the Icelandic **PROSE EDDA**

We are falling in the dark. Hunched over in a dimly lit little sphere, cold and cramped, we are descending from sunlit Earth into an unfriendly darkness nearly two miles below the Atlantic. Our purpose: to explore the planet's least-known mountain range, the Mid-Ocean Ridge. If all goes well, the chirping electronic gadgets around us, with our own observations, will give us new information about the forces that shape the Earth.

Like the Midgard Serpent of Norse legend, the Mid-Ocean Ridge rings the globe. It stretches through all the major ocean basins of the world, traversing some 74,000 kilometers (46,000 mi), covering nearly a quarter of the Earth, surfacing at scattered islands, and building the large volcanic island of Iceland. The Ridge is the largest geologic feature on our planet. The Himalayas, Rockies, Andes, and Alps could all fit end-to-end along it.

The Ridge is a world unto itself, populated by unique communities of animals that live in eternal darkness, supplied with oxygen by currents from the polar seas, bathed by thermal springs, and fed by the minerals of the living Earth. It is also a unique laboratory, where we can study the creative forces of the planet at work. On the basis of what we see there and the theories we formulate and test, we can come to understand better *how* the planet works. That's satisfying in itself—and may bring humankind enormous practical advantages as well.

The foundations of the sea

As explorers charted the seven seas and determined their large but finite dimensions, fear of the sea turned into curiosity. Voyagers ventured away from the shore to study the sea's mysteries, to sort out fact from fiction. Seventeenth-century scientists analyzed seawater, speculated on its saltiness, and pondered the tides.

In the 18th century, governmental hydrographic bureaus in Europe and the United States sent ships to probe the sea with "cannonball soundings." Crews lowered a cannonball tied to a rope to measure —often inaccurately—the ocean's depth. A sticky piece of tallow in a sampling tube carried down by the cannonball brought back a tiny quantity of bottom material. Its composition could then be marked on the charts.

Depth-sounding equipment became more elaborate and efficient in the middle of the 19th century, and explorers added measurements of water temperatures and currents to the data collected. They used steam winches, and sounding machines with detachable weights, to bring up seafloor samples from greater depths. These meager specimens provided an early inkling of the complex diversity of ocean life: Under the microscope they proved to be remains of living things, evidence of a vast marine graveyard.

One of these early oceanographic pioneers was Matthew F. Maury, *(continued on page 110)*

The Mid-Ocean Ridge: New Findings Show a Varied Face

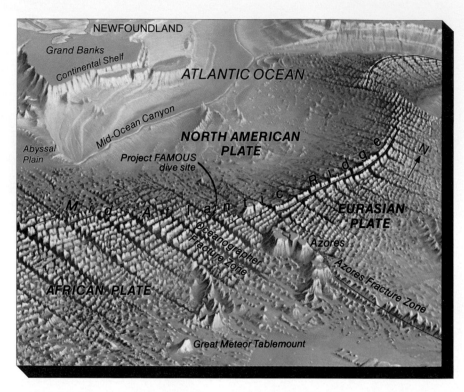

NEWFOUNDLAND

Grand Banks

Continental Shelf

ATLANTIC OCEAN

Mid-Ocean Canyon

NORTH AMERICAN PLATE

Abyssal Plain

Project FAMOUS dive site

N

Mid-Atlantic Ridge

Oceanographer Fracture Zone

EURASIAN PLATE

Azores

Azores Fracture Zone

AFRICAN PLATE

Great Meteor Tablemount

A wonder of the world hidden in the depths of the sea, the Mid-Atlantic Ridge and its central rift (above) dominate the center of the Atlantic seabed. This great scar forms as the North American Plate separates from the African and Eurasian Plates, a process that still continues. Columbus's voyage would take him a little longer now than it did in 1492: The Atlantic is about 10 m (30 ft) wider.

OVERLEAF: *The Mid-Ocean Ridge, some 74,000 km (46,000 mi) long, courses the Earth in this special projection designed to portray it as a whole. Though explorers have viewed only about 130 km (80 mi) of the Ridge, their instruments have probed it extensively. We now know that the Ridge, once pictured as a uniform chain of jagged mountains, is much more diverse —high peaks in the Atlantic; smoother, lower, and wider ranges in the Pacific; twisting terrain, abruptly changing from rough to smooth, under the Indian Ocean.*

The rate of seafloor spreading shapes this topography. Fast spreading makes wide, gentle slopes; a slower rate, high, rugged terrain. As newly created ocean bottom moves away from the Ridge, Earth's spherical geometry causes the rock to crack. Transform faults cutting across the Ridge are the result. The cooling, moving seabed contracts, subsides, and accumulates deepening layers of sediment, at last becoming the abyssal plain. Deep trenches mark the lines where two plates meet and one turns downward to be reabsorbed into the Earth. The Ninetyeast Ridge, its origin debated, arrows through the Indian Ocean; superimposed on the United States, it would form a barrier up to 4 km high from New York to Seattle.

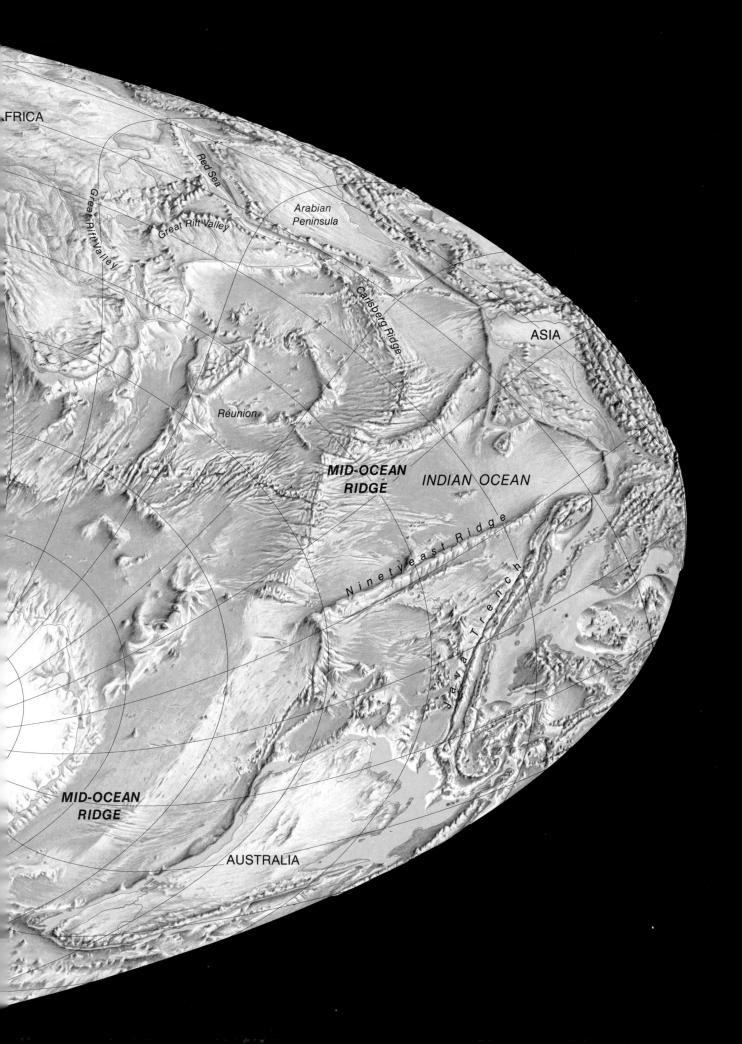

Sheltered between the twin hulls of its mother ship Lulu, Alvin *prepares to dive. Crewmen steady the sub as a giant cradle lowers it to the water.* Alvin *has been at work since 1964, and made some 1,100 dives over a mile deep in the first part of its career. In 1966 it located a hydrogen bomb lost from a U. S. aircraft off Spain. Today,*

with its depth limit extended to more than two miles, Alvin *makes about 100 dives a year for science, frequently in search of data on the Mid-Ocean Ridge.*

Superintendent of the United States Depot of Charts and Instruments. In 1854 he published the first bathymetric map of the North Atlantic Basin and followed it with his classic, *The Physical Geography of the Sea.* He painted a somewhat bizarre picture of the deep—"'a thousand fearful wrecks,' with that array of dead men's skulls, great anchors, heaps of pearl and inestimable stones...." But he also saw the grandeur. "The wonders of the sea are as marvelous as the glories of the heavens," he wrote. "Could the waters of the Atlantic be drawn off ... it would present a scene the most rugged, grand, and imposing. The very ribs of the solid earth, with the foundations of the sea, would be brought to light...."

Maury's style, along with his willingness to speculate, brought him a large international audience, and the huge hoard of data he accumulated proved useful to science and industry as well as ocean trade. His soundings of the North Atlantic particularly interested the founders of the Atlantic Telegraph Company, who intended to span the ocean with an undersea cable.

Maury pointed out what he called the Dolphin Rise or the Middle Ground, an elevated region on the mid-Atlantic floor "which seems to have been placed there for the purpose of holding the wires of a submarine telegraph." This was the first recognition of the Atlantic portion of the Mid-Ocean Ridge.

Laying the cornerstone of a modern science

In 1872, a great oceanographic expedition launched an exploration that was to alter permanently our understanding of the oceans. For four years, sailing over 127,000 kilometers (80,000 mi), a team of marine scientists circumnavigated the globe on Britain's H.M.S. *Challenger,* dredging, sounding, and sampling the depths of the sea. The treasure chest of new data, published in 50 stout volumes, became the foundation of modern oceanography. *Challenger*'s line soundings revealed what scientists later recognized as extensive undersea ridges, not only in the Atlantic but in the Pacific and Indian Oceans as well.

Between the World Wars, Germany sent the research vessel *Meteor* zigzagging across the North and South Atlantic in what proved to be another important scientific expedition. The ship had aboard new electronic echo sounders, the forerunners of modern sonar. The gear sent a single "ping" of sound downward, and the time it took for the echo to return revealed the depth of the ocean at that point. The *Meteor* expedition's new charts were easily the best of their time.

During World War II, the infant science of oceanography proved its worth to the United States. With an instrument called a bathythermograph, for example, scientists studied sea temperatures and the way thermal layering of the water could bend or even reflect echo-sounding beams. With this knowledge, Allied destroyers could use sonar more effectively to locate German U-boats. Oceanography emerged from the war a strong and growing science; vessels of many nations busily surveyed the seas, taking soundings across the ocean. Now continuously pinging echo sounders produced unbroken profiles of the bottom terrain, including the rugged Mid-Ocean Ridge.

Despite the long history of ocean-bottom soundings, it was not until the late 1950s that the Mid-Ocean Ridge was recognized as a global feature. And it wasn't until 1973 that humans would first lay eyes on the Ridge—four years *after* Neil Armstrong walked on the moon.

Testing the theory in the deep

By the early 1970s, geologists had accepted plate tectonics as a reasonable theory, but one that needed practical study. One proposal was to explore and map part of the Mid-Ocean Ridge in detail from deep-sea submersibles—diving vessels that many earth scientists then considered technological toys—in order to judge whether the theory fitted the facts: the "ground truth." Plate tectonics explained the Ridge itself as a spreading center where Earth crust was created, but we did not know exactly how that happened.

That proposal became Project FAMOUS (French-American Mid-Ocean Undersea Study), and I was an eager participant. We picked a site on the Mid-Atlantic Ridge some 650 kilometers (400 mi) southwest of the Azores that seemed typical of the worldwide mountain range. The great rift along the Ridge's crest is narrower here than in the African Rift valleys, apparently in proportion to crustal thickness. It measures 3 to 30 kilometers wide and 1.5 kilometers from rim to floor—a feature on the scale of the Grand Canyon—and marks the boundary between the African and North American Plates. They are separating at some two centimeters a year.

Like astronauts readying for the

Flying fish skim an oceanographic portrait, a 550-km cutaway across the eastern slopes of the Mid-Atlantic Ridge. The warm North Atlantic Current, fed by the Gulf Stream, supports surface life. Myriad tiny animals form the "deep scattering layer," so dense it confused early ocean soundings by reflecting sonar pulses. At night this living layer migrates toward the surface to feed. Below 1,000 m (3,300 ft), the deep sea's eternal darkness begins. Only 5 percent of the world's marine animals live here, most of them dependent on organic material that drifts down from sunlit levels. From about 1,000 m to 1,500 m, the water holds too little oxygen for some life forms to survive. Pressure and temperature at 4,000 m gradually dissolve seashells after the death of the creatures inside. Brittle stars and sluglike giant holothurians the size of cats live under pressure 500 times that at sea level. The North Atlantic Deep Water, a cold current inching from Arctic to Antarctic, brings them oxygen.

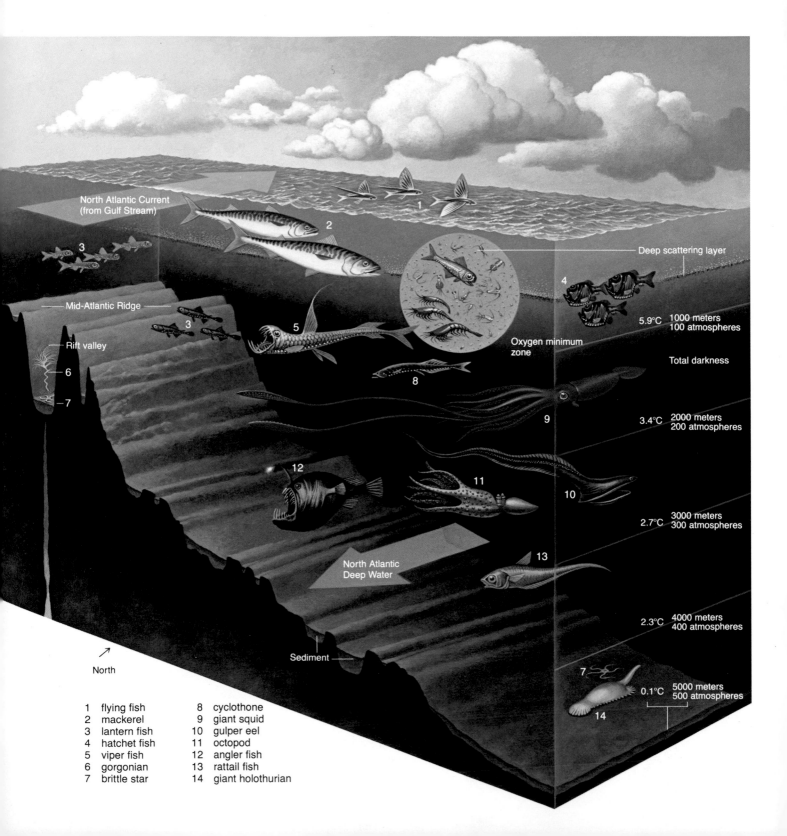

North Atlantic Current (from Gulf Stream)

Mid-Atlantic Ridge

Rift valley

North

Deep scattering layer

Oxygen minimum zone

North Atlantic Deep Water

Sediment

5.9°C 1000 meters 100 atmospheres

Total darkness

3.4°C 2000 meters 200 atmospheres

2.7°C 3000 meters 300 atmospheres

2.3°C 4000 meters 400 atmospheres

0.1°C 5000 meters 500 atmospheres

1	flying fish	
2	mackerel	
3	lantern fish	
4	hatchet fish	
5	viper fish	
6	gorgonian	
7	brittle star	
8	cyclothone	
9	giant squid	
10	gulper eel	
11	octopod	
12	angler fish	
13	rattail fish	
14	giant holothurian	

OVERLEAF: *Cruising along the edge of knowledge, French submersible* Cyana *(foreground) and bathyscaphe* Archimède *explore the Mid-Atlantic Ridge's* FAMOUS *section, 2.7 km (1.7 mi) down. Rattail fish investigate. An artist's brush brings light to the endless night of this rift valley, which explorers see only in floodlit patches. Behind* Archimède *the valley of a transform fault crosses the rift. Pillow-lava hills like Mount Venus (center foreground) mark the centerline of the rift, where lava wells up to create new seafloor. Fissures form as the North American Plate (left) and the African Plate separate.*

moon, the Ridge aquanauts spent many months in careful training. We would have only a few hours on the ocean floor, and we needed to be prepared to recognize and understand what we found. Scientists on our diving team went to Iceland and Africa to see rifting on dry land, and to Hawaii to walk on freshly cooled lava. We scuba dived to underwater flows that we thought probably resembled the Ridge's valley floor.

On the Ridge, about 2.7 kilometers down, we would be mapping and observing objects from a few meters in size down to mere centimeters. But available maps showed features smaller than 100 meters only vaguely—or not at all. So as we trained, other scientists from France, Britain, Canada, and the United States conducted more than 25 cruises to the FAMOUS site. They photographed our target area and crisscrossed it with seismic, magnetic, and gravity- and heat-sensing equipment. The U.S. Navy surveyed it with an advanced sonar system that produced superbly detailed ocean floor maps.

To inspect that terrain, we would use two different types of diving craft: a bathyscaphe and two submersibles. The bathyscaphe, invented by Swiss physicist Auguste Piccard in the 1940s, works like a subsurface balloon. But instead of hanging from a bag full of lighter-than-air gas, the gondola—a metal sphere strong enough to resist the enormous water pressure of the deep—is supported by a float of lighter-than-water gasoline. In order to sink, you valve off gasoline from the float; to rise, you drop ballast—a stream of iron shot.

Big and clumsy, the bathyscaphe

is hardly more maneuverable than an elevator, but it can dive to great depths, and in 1960 Piccard's *Trieste* made the world-record 10,915-meter (35,810-ft) descent to the lowest point on Earth: the Mariana Trench off Guam. Nonetheless, bathyscaphes are diving dinosaurs now, and though the French *Archimède* made our preliminary dives, bathyscaphes can't compete with modern deep submersibles like the French *Cyana* and American *Alvin*, mainstays of Project FAMOUS.

Jeeps of the deep

These miniature submarines use a new flotation material called syntactic foam. Composed of glass microspheres mixed with epoxy, the foam can resist deep sea pressures but is much lighter than gasoline, so less of it is needed to float the heavy steel or titanium pressure spheres. As a result, the craft are less than a third as long as the bathyscaphes: *Alvin* measures just 7.6 meters (25 ft). Instead of being slowly towed to dive sites, the submersibles can ride on the deck of a mother ship. They are also very maneuverable—good for work along steep underwater slopes and in tight corners.

By the summer of 1974, all the preliminary work was finished. The American diving program in the rift valley could begin in earnest. My greatest fear was not the danger of diving, for *Alvin* is probably the safest deep submersible in the world. I was afraid I would not understand the geological significance of what I was about to see.

Project FAMOUS was a complex and expensive undertaking, and the first use of a submersible for mapping at great depths. The FAMOUS

team had advocated the project even though some eminent scientists doubted that it would produce useful results. If we weren't able to relate what we saw through *Alvin*'s viewports to the theory of plate tectonics, then the doubters would be proved right. We would be nosing around in the dark with little lights, hoping to reach a better understanding of the Earth as a whole.

To the mountains of eternal night

Our dive begins early in the morning, as all of them will this long exciting summer. While the crew readies the submersible, the science party directs *Alvin*'s mother ship, named *Lulu*, into position for launch. Our target is the low ground between two hills that we have named Mount Venus and Mount Pluto, in the very center of the rift valley. A network of acoustic beacons was dropped into position several days ago, so by sending out a ping from *Lulu*, we can determine our exact position by timing the beacons' answering pings.

By nine o'clock the pre-dive checks are finished and the launch begins. Two pilots and I scramble aboard. Normally *Alvin* carries two scientists and one pilot, but this is our first dive to this rugged terrain, where *Archimède* reported strong currents last year. We don't want to take chances. Once free from *Lulu*, the pilot makes his final checks. "*Lulu*, this is *Alvin*. My hatch is shut, blowers are on, leaks and grounds are normal, no joy [echo] on the fathometer, oxygen is on, tracking pinger on. Ready for dive, request permission to dive. Over."

"*Alvin*, this is *Lulu*. You have permission. Dive when ready."

Alvin's instrument-laden bow pokes into a chasm in a field of pillow lava on the Mid-Atlantic Ridge. Seawater filters down such fissures toward hot magma. In the crevice lives a crinoid, kin to starfish. Tendrils of gorgonian coral (right) festoon pillows on the Ridge in a photograph taken through Alvin's porthole.

OVERLEAF: Alvin, *escorted by fish heading for a baited camera far below, nears a draining lava lake along the East Pacific Rise, the fastest-spreading section of the Mid-Ocean Ridge yet found. In this artist's conception of a future dive, transform faults offset segments of the rise behind the dark plume of a forceful "black smoker" hot spring.*

The long, dark fall

We leave behind the roll of the ship and the noise of feverish activity on deck; *Alvin* begins to fall toward the ocean floor. At first the air inside the submersible is hot and humid; I rest quietly while the pilot rechecks the interior. We can still see some light from the sunlit surface; around us the sea abounds with the microscopic plants on which all other ocean life depends.

The light in our viewports grows dim as we sink 30 meters a minute. At 365 meters (1,200 ft) the sunlight is gone, and only our cabin lights relieve the darkness. The 2,740-meter descent—a 1.7-mile drop—will take an hour and a half. Gradually the temperature drops too, chilling the humid air. Beads of moisture condense on the cold walls of the tiny pressure sphere. *Alvin* is unheated, its battery power needed for more important uses than comfort.

Now the marine life we see looks less and less like the creatures we know from the surface. Outside our viewports cold sparks flash as the bioluminescent inhabitants of these depths, responding to *Alvin*'s sudden bulk and dimly lighted viewports, turn on their own live lights. Sifting down past the living are the waste matter and the dead from high above. We are falling faster than this macabre marine snow, so it slips upward past the viewports, providing our only sense of motion.

At last a noisy pinging fills the sphere; our sonar has picked up the walls of the valley. As the bottom comes up at us, the pilot switches on the outside lights and drops ballast weights to arrest our fall. We hang motionless in neutral buoyancy. Outside, 250 atmospheres of pressure—more than 1.5 tons per square inch—squeeze *Alvin* in an invisible fist.

"*Lulu,* this is *Alvin.* My altitude is 30 meters. I am driving to bottom. I'll contact you when we land."

On the edge of creation

Slowly, the seafloor comes into view. Glassy surfaces reflect our lights as we touch down. Unseen to the north lies Mount Venus, to the south Mount Pluto. Little sediment coats the bare rock, but the scene is not devoid of life. *Alvin*'s floodlights pick out a sea anemone here, a whip coral there, bright in the first light ever to strike them. Their kind lived on these slopes in total darkness long before the pharaohs ruled in Egypt, and when our dives are done this summer, they will never see light again.

But it is the terrain itself that interests us. As we observe, map, photograph, and collect samples—never able to see beyond the 30-meter (100-ft) reach of *Alvin*'s lights—the realization grows that the rift valley landscape has been created by volcanic forces, not just shaped by tectonic ones. We expected to find mostly tectonic features, like faults, that result from rock movement; the theory we are testing is called plate tectonics, after all. But though we do see such forms, and though tectonic forces are moving the plates apart, tectonic action can't fully explain this landscape-in-creation. The rock is lava, and most of the landforms we see are created by very recent volcanism.

Since we are dealing with crustal creation, we are most interested in the freshest lava, which drapes the valley slopes in bulbous, glassy

Deep-sea creatures—varieties of coral and a stalked crinoid—find a lifelong niche on a lava pillar on the East Pacific Rise near the mouth of the Gulf of California. The ribbed, hollow pillars form in molten lava lakes: Magma surges up through fissures and floods rapidly across the seafloor, trapping water in rock crevices.

Superheated, this water expands and blasts through the molten magma above, which hardens into a tube around the jet. When the lake later drains (preceding pages), the lava leaves a series of ledges like bathtub rings around the pillars and the edges of the lake.

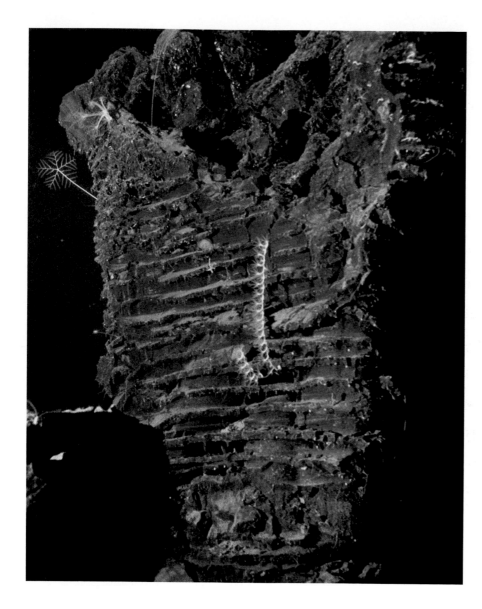

flows called pillows, the unique form that lava takes when it flows underwater. As lava makes contact with the near-freezing seawater of the valley, the lava's surface quickly solidifies into a bulge of black, glassy basalt. The interior, still molten and still under pressure, bursts open the pillows or stretches them to build bolster and tube shapes, secondary bulges, and toothpaste-like extrusions.

When we maneuver *Alvin* up the southern slope of Mount Venus, we see it is one massive pile of pillow lava, ornamented with glassy buds that testify to the lava's freshness. I try to imagine the scene when Mount Venus last erupted: No red-hot molten streams like those on land, only small red cracks, gleaming briefly, then winking out. No escaping steam or gas, just rock as stiff as fresh concrete, slowly and irresistibly squeezing out against the water and rapidly hardening in the 2°C temperature to form lobe on lobe, pillow on pillow of lava, until oozing lava had built a 250-meter hill in the center of the rift and paved the floor of the valley with rounded shapes.

These young lava flows do not stay unaltered long—geologically speaking, at least. Within a thousand years after they form, seawater has etched their glassy surfaces, the buds have broken off, and settling sediment has dusted them. More important, the great conveyor belts, the separating plates, have begun to pull them apart, making room for still newer lava crust. Deep fissures appear, most of them running parallel to the axis of the valley. We drive along one such crack for over 100 meters before it

Pressure and cold water shape pillow lava—the most common lava formation on Earth. Piles of pillows make up much of the new ocean floor created in spreading zones. The pillows often build hills as high as 100 m. Natural fracturing may reveal solid centers like these, or hollow ones if the pillow's core has drained away.

peters out. Not until later do we realize that these fissures are another piece in the plate tectonics puzzle.

The case of the missing heat
About the time that Project FAMOUS was under way, other scientists were working on a problem halfway around the world. A few years before, researchers had shown that the Mid-Ocean Ridge is a ridge partly because the young rock of its interior is swollen by tremendous heat. As the plates move apart, the rock cools, contracts, and subsides. By the time the crust has aged about 80 million years, it has lost the original magma's excess heat and—now carried some distance from the Ridge where it was formed—rests at deeper ocean-basin depths.

Since we know how hot the lava forming the Ridge was originally, and since we know how many millions of years it takes to cool completely, we should be able to calculate how much heat any particular point on the Ridge is giving off as it cools. That was the theory, at least. But when scientists dropped heat probes onto the outer slopes of the Ridge, they found that the rock was not hot enough. The Ridge was losing heat somewhere else—but how?

One idea was that cold seawater might be penetrating the crust, picking up heat from the rock and bringing it back up to the seafloor; then the heat measurements would make sense. Drill cores provided some evidence for such circulation, and the many cracks and fissures we had seen during Project FAMOUS could provide the path for seawater to enter the new crust and percolate down toward the hot magma chamber below. We had even seen some

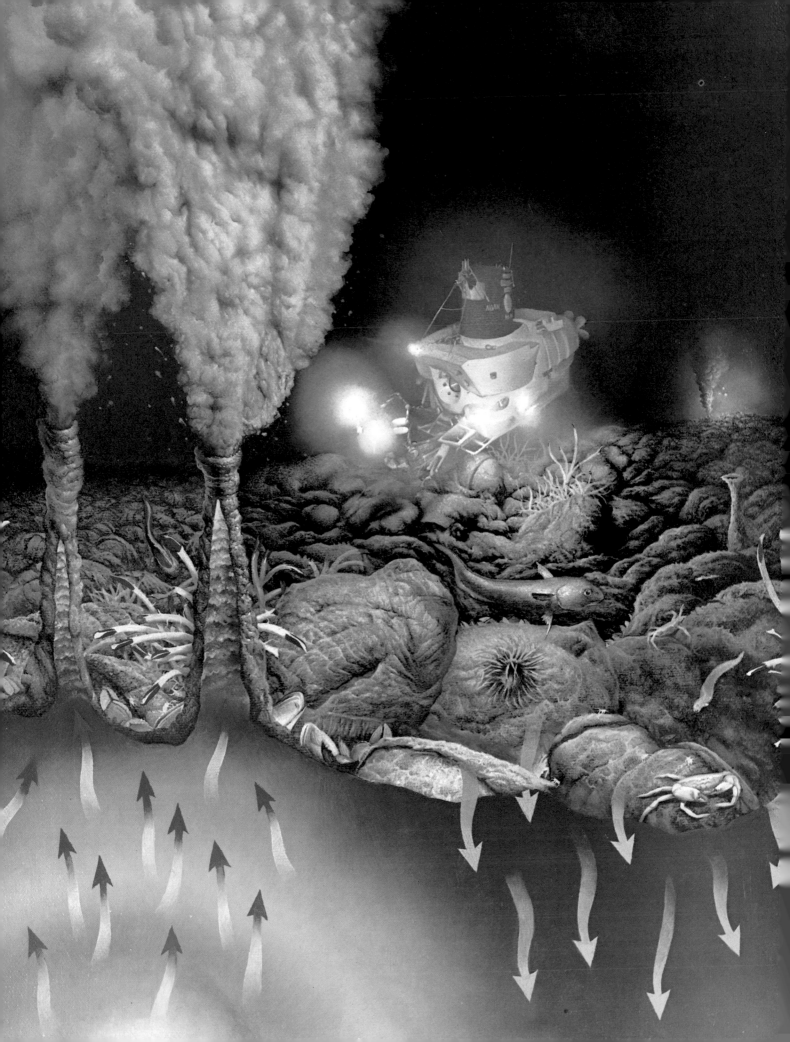

Alvin sheds light on Earth's only complex ecosystem not based on photosynthesis. Minerals and heat, not sunlight, fuel these seafloor oases. Seawater (blue arrows) seeps down to magma beneath the Ridge, becomes superheated and saturated with minerals, and then rises (red arrows), building natural chimneys. Bacteria flourish in the warmth by metabolizing hydrogen sulfide from the vent water. They form the base of the food chain for such oasis creatures as the eel-like "21-degrees-north vent fish." Giant clams, tube worms, rattail fish, anemones, and crabs cluster around this fountain of life. Beyond, a lighted camera sled examines another oasis.

signs of hydrothermal action—but no one had ever seen an active vent returning the heated seawater.

A new world on the Galápagos Rift
If water was really flowing in and out of the Earth's interior at the Ridge, carrying heat away, we reasoned, then there must be thermal springs somewhere along the summit of the Ridge. From earlier studies, we decided the place to look was on the Galápagos Rift, a branch of the East Pacific Rise.

There, teams of scientists from the Scripps Institution of Oceanography, Oregon State University, and Woods Hole Oceanographic Institution had cruised the waters, towing electronic thermometers just above the valley floor. To their delight, they found small areas where the water temperature was a fraction of a degree higher than the surrounding, near-freezing levels.

A water-sampling bottle, opened by remote control when the temperature jumped, recovered forms of dissolved helium and radon gas that could only be coming from deep within the Earth. Gas and temperature together were solid evidence that the springs were there. We had to take a look at them.

So the Galápagos Hydrothermal Expedition of 1977 set out. From our staging area at the Panama Canal, we sailed to the dive site on the research vessel *Knorr*, ahead of *Lulu* and *Alvin*. Once there, we set up the network of seafloor beacons we would use to track *Alvin* and our deep-towed camera system, *Angus*. We would use *Angus* to find the springs and *Alvin* to investigate.

Angus, carrying a temperature sensor, was lowered on its cable, and the vigil began. For hours, we towed the camera sled back and forth above the rugged lava below. The water was nearly 2,500 meters (8,200 ft) deep and we had to keep *Angus*'s cameras within five meters of the uneven bottom. Watching the sonar signals from *Angus*, we called instructions to the winch operator through the night. "Up five! Down five! Up, *up!*" Now and then the sled would slam into a rock cliff below, sometimes snagging itself. The tension would rise on the cable—and on deck—until *Angus* broke free or *Knorr* backed down to keep the towing cable from breaking.

Finally, the temperature trace spiked across the plotting paper and dropped back again: *Angus* had passed through warmer water. Our spirits jumped as well. Our eyes fixed on the plotter, waiting for another jump. Minutes went by—then an hour. After six hours *Angus*, out of film, had to be pulled up.

After our shipboard darkroom had processed the film, we quickly reeled through the 3,000 frames, most showing barren lava, to find the pictures shot as *Angus* passed through the temperature jump. We weren't sure what the spring would look like, but we hardly expected to see what we found: Photographed through water turned strangely milky, a bed of giant white clams flourished 2.5 kilometers down.

Alvin dived to investigate, and the voices coming from below were filled with excitement. Later I was able to go down myself. After many years, deep diving had become routine for me—but not that day. Surrounding the vents of warm water were incredible concentrations of life: foot-long clams, giant tube worms, crabs, mussels, and fish. We had found a whole new, luxuriant world of creatures somehow living comfortably in unimaginably hostile conditions.

The secret of the oases
After the initial excitement of discovery wore off, we began to wish we had a biologist aboard. How could all these animals live in a region where no light penetrates? Every complex ecosystem we knew of depended on converting the energy of sunlight into food. Without sunlight, there is little food—and little life. On the dark seafloor there can be no photosynthesis, and so the deep ocean—normally—is a virtual desert. On our earlier dives at the Mid-Atlantic Ridge, we had seen only an occasional sponge, coral, or fish, all dependent on organic matter drifting down from above. But here the bottom of the sea teemed with life that totally covered some of the lava flows. What was feeding this oasis?

The answer came in the laboratory aboard ship, when we opened water samples taken at the vents. A foul, sulfurous odor filled the lab, so strong we had to throw open the portholes. The rotten-egg stench proved to be, as one scientist put it, "the sweet smell of success."

Just as springs in a desert oasis supply the water that the desert lacks, the springs here supply what the undersea desert lacks: energy. The gas we smelled was hydrogen sulfide, and water from the vent later proved to contain high concentrations of bacteria—more than 250 types. (The gas and the bacteria, along with the water's heat, produced the milky look that had

123

Life in a Sunless World

Alvin's *temperature probe*
registers a normal—but chilly—
2°C as it tests the water on a dive
to the Galápagos Rift. Away from
the warmth of a nearby vent, an
anemone more than 58 cm (23 in)
across shares a ledge with a brittle
star, a crab, a spiky protozoan, and
a nameless rosette-like creature.

OVERLEAF: *In the vent's warmth,*
foot-long clams crowd the rocks.
They live by filtering bacteria—a
billion in every liter—from the
warm water. Growing 500 times
as fast as related species, they
can mature and reproduce
before the spring dies out.
Unlike typical clams, these have
red flesh rich in hemoglobin,
an adaptation to the low
oxygen levels in the vent water.

2ND OVERLEAF: *Red-plumed tube*
worms, some more than 3.5 m
(12 ft) high, thrive near a vent. The
worms live in symbiosis with vent
bacteria that pack their eyeless,
mouthless, gutless bodies. Using
oxygen that the worms absorb
through their plumes, the bacteria
break down hydrogen sulfide to
make food for their hosts.

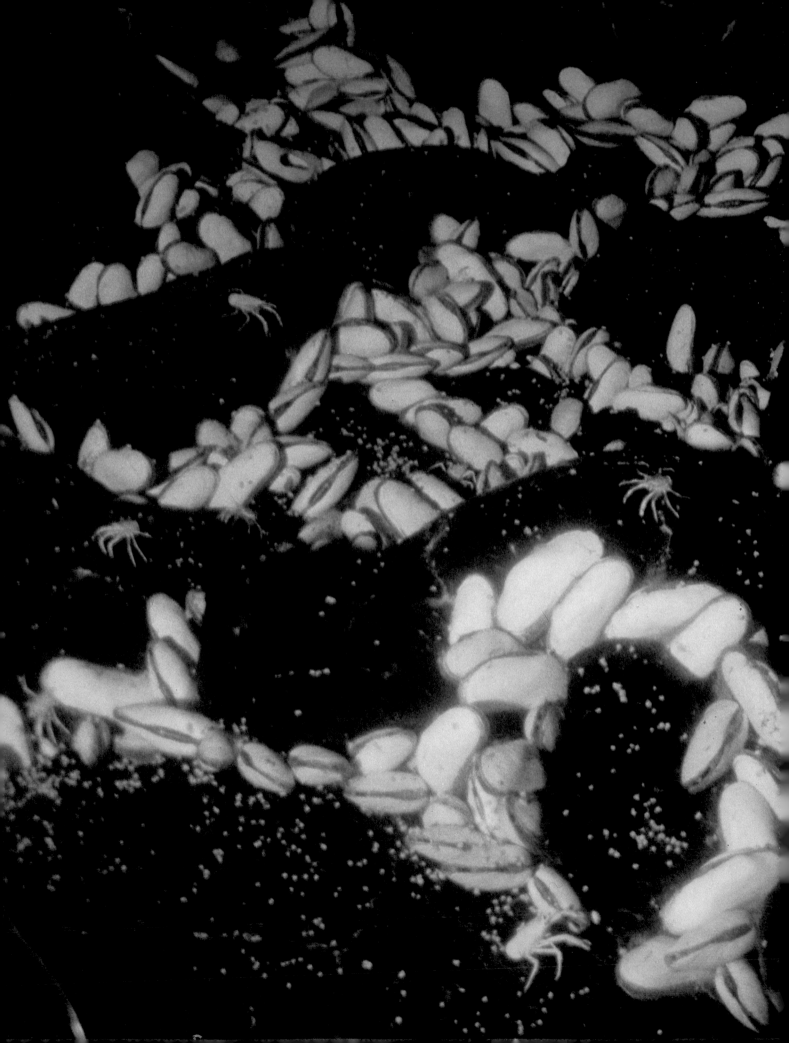

puzzled us before.) The bacteria apparently live in the creviced, porous lava, nourished by the flow of vent water. Unlike life that relies on the sun's energy, the vent bacteria use energy liberated by combining the oxygen in seawater with hydrogen sulfide. With this energy, they convert inorganic carbon dioxide in the seawater into the organic carbon compounds necessary for life.

When biologists later examined clams and mussels that *Alvin* collected at the vents they found rich concentrations of bacteria in the creatures' stomachs. The bacteria are the base of the oasis food chain: The large creatures there either eat the bacteria, live in symbiosis with them (like the giant tube worms), or eat the animals that live on the bacteria. We had found the first complex ecosystem based not on photosynthesis, but on chemosynthesis.

Many of the animals of the warm springs are new to science, and in fact have been classified as entirely new families. The discovery of a community independent of the sun has also interested scientists investigating the origin of life.

We suspect that seafloor rifts and underwater thermal springs have been around for 2 to 3.5 billion years. We now know that these sites are friendly to life and, in fact, provide an ample source of energy for the living communities that surround them. When the phenomenon of rifting was new, and when the oceans formed by rifting were still shallow, rift springs could have provided an excellent environment for life's first development from lifeless molecules.

In 1977, however, we thought the springs in the Galápagos Rift might be a freak of nature. So far they were unique, and one-time phenomena make scientists uneasy. But just two years later a new discovery ended that worry.

The smelters of the deep

Work on the East Pacific Rise off Mexico, at latitude 21°N, revealed more thermal springs in 1979. This time when we dived we found not only a similar biological community of bacteria and animals, but also "black smokers," tall stony chimneys in the middle of the oasis, with opaque clouds of black particles jetting straight up from them. They looked like the steel plants of yesteryear, and the updraft of water gently pulled *Alvin* closer.

Luckily, we were able to keep our distance, for when we inserted a temperature probe in one chimney, the instrument melted, though it was made of the same tough plastic as *Alvin*'s viewports. Later measurements revealed the jet of water to be 350°C, hot enough to melt lead. The chimneys themselves proved to be solid minerals deposited by the hot water: iron, zinc, copper sulfides, and a little silver.

Scientists have yet to assess the full impact of these discoveries, but they wonder if the drifts of minerals around the underwater vents might be commercially exploitable ore deposits. Mining companies have already become interested, and nations have begun to debate ownership of the minerals. We geologists are still trying to work out the conditions under which smokers form, how to predict where they will deposit which minerals, and how long such an energetic phenomenon as a black smoker might last.

Salt and the second cycle

The springs and smokers have already solved one problem for us, though. For years, chemists had tried to unravel an oceanographic mystery: How did the sea acquire its chemical composition, including its salt? They had thought the answer lay in rivers, which carry minerals leached from rock to the sea. But water samples showed that inflow from rivers could not have produced the chemistry of the oceans. The proportions were wrong, with some elements too plentiful, others too scarce.

The black smokers and warm springs of the Ridge explain much of this mystery. Now we know that the world's water circulates in at least two ways. The first is the one children learn: Clouds drop rain on the land; from rock and soil, flowing rainwater picks up minerals that rivers carry to the sea; the seawater evaporates, leaving the minerals behind, and forms clouds again.

The second cycle is the newly discovered circulation of seawater through the crust of the Mid-Ocean Ridge and back. Heated deep in the rock, the water deposits some of the minerals dissolved in it. The rock in turn releases other minerals and heats the seawater into a hydrothermal brew containing hydrogen sulfide and dissolved heavy metals. Loaded with these ore-makers, the water rises again and escapes, seeping from the thermal springs or pouring out the chimneys.

Many scientists now agree with marine chemist John Edmond's estimate that the whole volume of the oceans circulates through the crucible of the Mid-Ocean Ridge every eight to ten million years, growing poorer in some elements, richer in others. (The rain-and-river cycle, by contrast, handles that volume 200 times as fast, but carries only $1/100$ to $1/1000$ of the concentrations of chemicals.) So even if the Mid-Ocean Ridge is only about 2 billion years old, the oceans must have flowed through the Ridge at least 200 times. They owe their chemical composition as much to the Ridge as to the rivers of Earth.

But why have our dives found thermal springs and black smokers in the Pacific Ocean and not in the Atlantic? We've seen such a small part of the seafloor that the difference could be only a coincidence. More likely it's because the two ridges are spreading at different speeds. The crust in the Atlantic grows about two centimeters a year, but at the Galápagos Rift and at 21°N on the East Pacific Rise the plates are separating three times faster. That means three times as much magma seeps from Earth's interior, providing much more heat for the water that finds a path below the seafloor. And there's good evidence that the magma chamber in the Pacific doesn't lie as deep as it does in the Atlantic, so the water's circulatory path is shorter.

We can test this explanation by looking at a site on the Mid-Ocean Ridge that is spreading still faster. We would expect more mineral deposits at such a site, and there would probably be other differences we haven't even thought of yet. On the East Pacific Rise, off Easter Island, where the Pacific and Nazca Plates are separating at a speed *nine* times that of the Mid-Atlantic Ridge . . . there we should find good hunting.

Mother ships like the U. S. Navy's Point Loma once tended the bathyscaphe Trieste II. Much smaller surface ships will serve as command posts for future undersea exploration. Robots with stereoscopic, color-television eyes and powerful floodlights will give human eyes longer looks below; camera sleds will photograph the seafloor in four-acre swaths. By guiding cameras from the mother ship, scientists can avoid the expensive, hours-long commute to the bottom. Even so, they face a formidable task; 99.8 percent of the Mid-Ocean Ridge remains unexplored today.

*district, despite new eruptions
in the area. Explosions built
Hverfjall, the 2,500-year-old
cone behind them, within a few
weeks, when rising magma flash-
heated groundwater into steam.*

Bump on the backbone of the Mid-Atlantic Ridge, Iceland offers a dry-land view—as tourist leaflets boast—of "continental drift in action." A forked volcanic zone divides the island where two plates part company. Since 1975 much of the action has occurred near Lake Mývatn's Krafla volcano (lower), where new fissures have opened every few months. Inflation of an underground magma chamber lifts ground level as much as a meter before an eruption. Lava deposited by the "Mývatn Fires" of the 18th century (red) presaged these new "Krafla Fires" of the 20th (pink). The 250-year gap suggests that Earth's plates do not separate smoothly, but in episodic jerks.

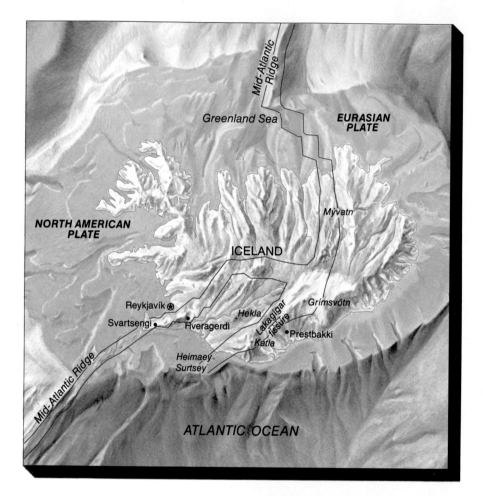

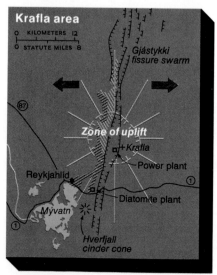

The Midgard Serpent will blow so much poison that the whole sky and sea will be spattered with it. . . . PROSE EDDA

Our driver stopped the car in front of his home near Lake Mývatn. He pointed along the lakeshore to the inactive crater of Hverfjall, which he can see from his house.

"Do you see the other volcano just beyond Hverfjall's summit?" he asked me and my Icelandic colleague Haraldur Sigurdsson. "Last summer, I couldn't see that from here."

The ground on which his home stood had risen about a meter, just enough for him to see over Hverfjall to the next peak.

Three kilometers beneath us, we knew, molten magma was forcing its way up through passages in the Icelandic rock. Over the months it had inflated a subterranean chamber so that the land above rose like a car on a hydraulic lift.

It's one thing, I reflected, to see Earth come alive in exotic places far from home, quite another to watch the landscape rearrange itself in the window above the family TV.

I had come to this part of northern Iceland to catch mid-ocean spreading in the act. Here, fissures next to the volcano Krafla had been intermittently spouting lava fed from the magma chamber below us.

These new eruptions were adding one more chapter to the story of Iceland's cloven existence astride the Mid-Ocean Ridge. Part of the island moves westward on the North American Plate; the other part eastward on the Eurasian Plate. In between, molten rock makes its way

Standing between North American and Eurasian Plates, field observer Hjörtur Tryggvason braves winter ice to measure how much a fissure in the Krafla area widened overnight. Metal bars cemented to each side make a jury-rigged "riftmeter." During an eruption, fissures may jump in width by a meter. Lasers more accurately gauge the long-distance change across the rift zone as a whole. Krafla's stirrings have opened the rift 5 m in all: two and a half centuries of plate movement packed into just a few years.

to the surface, creating more land. Iceland's position on the Ridge is also above one of Earth's hotspots, which is probably why the island stays above water; volcanism is building it faster than spreading can carry the new rock away.

"The rate of spreading is about two centimeters a year," says my friend Haraldur. "The new crust in Iceland is added about as fast as your fingernails grow." But the growth is erratic. Scientists had once assumed that the seafloor was spreading at a steady pace along the entire length of the Ridge. Yet in Iceland, the only place we can study the Ridge above water, we were finding otherwise.

Vikings witnessed volcanic eruptions
What I came to see for scientific reasons, Icelanders have had to live with for generations. Vikings sailed the North Atlantic to colonize these lava-black shores in the ninth century, and saw eruptions that probably inspired the story of Ragnarök, the old Nordic version of doomsday. The volcano Hekla's frequent fiery outbursts led mariners of the time to call it the entrance to hell. The Norse settlers, however, accepted their island's violent moods, and succeeding generations have grown accustomed to an average of one eruption every five years.

Modern Icelanders harness volcanism's gift—geothermal energy—for a host of purposes: central heating and hot water, electrical power, horticulture, and swimming pools; they even tap volcanically heated seawater to extract packing salt for the fishing industry.

But Icelanders pay for living on the edge of creation, and never was

Volcanologists approach a searing spray of cinders during the 1973 Heimaey eruption. Iceland may be receiving a double dose of such ejecta; Icelandic magmas yield chemistry typical both of spreading ridges and of deep-seated volcanic hotspots. These two sources might account for Iceland's output—a fifth of Earth's recorded lava flow.

Lakagígar, a lethal row of craters up to 90 m (300 ft) high, sleeps under blankets of lichen and moss. In 1783 this great fissure emitted a world record volume of lava.
OVERLEAF: *Deadly volcanic haze veils the sun as Lakagígar's molten tide swamps farmsteads. Blighted pastures poisoned livestock across the country, first omen of famine.*

the price higher than in 1783, when an eruption in the south almost forced evacuation of the entire country. These lava flows, Earth's largest in historical times, offer clues on how spreading occurs. So on my arrival in Reykjavík, Haraldur and I set out for the site of Iceland's grimmest hour—Lakagígar.

We followed one of the island's rare paved highways toward the southern coast, through open landscapes of barren rock, moss-covered heath, and rugged pastureland. By the time we could see the glacier capping the volcano Katla, our highway had turned into a standard Icelandic road, a rough but well-used track of gray-black lava grit.

Past Katla, amid a washed-out plain of sand and gravel, a pothole bent a wheel rim. While we put on the spare, Haraldur explained that when a glacier-covered volcano erupts, its ice cap suddenly melts. Within minutes the ensuing flood, called a *jökulhlaup,* or glacier burst, thunders out across an expanse like the one where our disabled car now sat. Katla eruptions have yielded discharges that briefly exceeded the flow of the Amazon, the world's most voluminous river.

Farther on we reached a very different, other-worldly plain. Jagged, jumbled rock, its raw edges softened only by moss, stretched toward the horizon—a huge lava field. In Iceland's cool climate a century may pass before even moss gets a grip on hardened lava. It takes centuries more for grass and then shrubs to take hold.

This lava field, Haraldur said, is only one portion of Lakagígar's incredible output. On the far side we came to some farms. They were full

of life; farmers were busy cutting hay, laying it out to dry, and baling it for the winter months. What a contrast, I thought, with the events recorded here by Reverend Jón Steingrímsson, parson of the stone and turf church at nearby Prestbakki 200 years ago.

The Midgard Serpent's fury

It is June 8, 1783. Few of Jón Steingrímsson's parishioners still worry about the earthquakes that rocked the region last week; earthquakes are not unusual in Iceland. But these quakes were a warning. At 9 a.m. huge fountains of lava spout from a giant fissure that opens in the barren lands to the north. Glowing rock begins to stream down riverbeds toward the farmlands below.

In the following weeks families watch their land and homes disappear beneath a simmering blanket. On July 20, as the lava flood rolls slowly toward the farmsteads of the Prestbakki area, Jón Steingrímsson summons his terrified congregation to the church and leads them in a prayer for deliverance. When the people file out, they discover that the lava has halted only a short walk away. Jón Steingrímsson's defiant "Fire Mass" will earn him a place in Icelandic legend, but Lakagígar is not yet done.

The eruption continues, from what is now a 25-kilometer-long row of at least 100 craters. A toxic bluish haze lingers over Iceland throughout the summer, stunting the grass that feeds sheep, cattle, and horses. What grass there is seems poisonous, and animals that graze on it begin to die.

Jón Steingrímsson keeps a detailed record of the disaster: "The

horses lost all flesh; on some the hide rotted all along the back; manes and tails decayed and came off at a sharp pull. . . . Those people who did not have sufficient old and wholesome food throughout this time of pestilence also suffered. . . . Their bodies puffed up; the gums swelled and cracked, with sore pains and toothache."

The eruption lasted eight months. By the end, the flows had covered 565 square kilometers—the equivalent of burying an area almost twice the size of Chicago three stories deep in lava. Yet no one died in this flood. Death came later, slowly.

The fissure had released staggering amounts of gas—50 million tons of sulfur dioxide (the same pollutant that helps create acid rain), 19 million tons of carbon dioxide, 5 million tons of fluorine—all part of the deadly blue haze. What probably killed the livestock was fluorine poisoning. Half of Iceland's cattle died, and over three-quarters of the horses and sheep.

The short, high-latitude growing season made recovery impossible. Food shortages and disease took a catastrophic toll. The Haze Famine, as it came to be called, eventually killed 10,000 people—a fifth of the country's small population.

Lakagígar's pollution spread far. It blackened grasslands in the Faroe Islands 500 kilometers (310 mi) to the southeast, gave Scotland the "year of the ashie," tarnished copper kettles in Kent, dropped dust in Italy, and spread haze to Siberia.

In Paris Benjamin Franklin, the plenipotentiary to France from the 13 new United States of America, speculated about the global effects of "a constant fog over all Europe and a great part of North America. This fog was of a permanent nature; it was dry and the rays of the sun seemed to have little effect. . . . Hence, perhaps the winter of 1783-4 was more severe than any that had happed for many years."

Haraldur has compiled 18th-century weather records that confirm Ben Franklin's suspicion: Lakagígar's haze did indeed lower temperatures. The following winter in the eastern United States was 4.8°C (8.6°F) degrees colder than normal, the greatest drop caused by any eruption from 1738 to the present.

Haraldur's research also leads him to conclude that in Lakagígar's plumbing system the magma had not come straight up from deep in the earth, but sideways, along underground cracks pulled open by rifting—cracks linked to a magma chamber 40 to 70 kilometers northeast, under the volcano Grímsvötn.

Did the current Krafla eruptions in northern Iceland work the same way, from fissures fed indirectly? If so, what did this pattern mean? We flew north to take a look.

The volcanoes of Midgewater

From our small plane I gazed down on Mývatn—Midgewater—a shallow lake set on a high plateau of the volcanic zone. To the north stood the low, broken peak of Krafla, remnant of at least 20 eruptions in the last few thousand years. Beyond, wisps of steam rose where thin sheets of fresh lava blackened uninhabited rangeland.

From the air I noticed numerous small, shallow craters framing the lake. They were not volcanic vents, Haraldur told me, but "pseudo-craters," formed long ago when a

Fountains higher than a 25-story building shoot from a system of fissures 5 km long at Krafla. After taking this picture from a low-flying plane, volcanologist Sigurdur Thórarinsson, veteran of many Icelandic eruptions, camped out near the fissure. "The view from the tent was marvelous, and it was warm and nice in spite of the

cold weather, but the sulfuric
stench from the lava was at times
rather bad. The rumbling gradually
made me sleepy." This eruption
lasted for five days in January
1981. Krafla-watchers learned to
predict such outbreaks, but only by
a few hours.

OVERLEAF: *Ghostly glow of
volcanic gas, ignited by magma
just below the surface, intrigues
scientists at Krafla. Such
"degassing" gave Earth its first
atmosphere. Gases released four
billion years ago did not burn,
since there was no free oxygen.*

Earth's heat combats air's chill in daily Icelandic life. Despite a cool climate and icy rivers, hot-spring-fed pools have turned swimming into a national pastime even in January (opposite). In Hveragerdi, natural steam heats greenhouses (below) that grow not only needed vegetables, but flowers—highly valued in this bleak land. Red-shaded sunlamps lengthen the subarctic's short winter day. Hveragerdi baker at a geothermal-steam oven (bottom) prepares brown-black hverabraud, hot-spring bread. Icelanders consider the steam-moisturized loaf a delicacy, served hot with melting butter and mutton pâté or herring.

Iceland's answer to the energy crisis: A geothermal plant at Svartsengi not only heats seven towns and a NATO base but generates electric power as well. Pressure-separator towers feed steam to a turbine and to heat exchangers, which warm the water piped to consumers. To equal this plant's top output, its customers would have to burn 100 barrels of oil an hour. Runoff from briny steam, tapped as far as a mile down, leaves a glistening mineral residue. Geothermal wells, or boreholes, provide 75 percent of the country's population with inexpensive central heating.

lava flow flooded the waterlogged land. Trapped water quickly turned to steam and exploded through the layer above like popping bubbles, leaving the pseudocraters.

The Mývatn area abounds in such geological curiosities—fissures that hold crystal blue swimming holes fed by hot springs, a lava flow that collapsed into a tangle of twisted and tortured rock called the Black Castles, and a bubbling geothermal basin beneath a yellow hillside whose sulfur helped make gunpowder for medieval Danish weapons.

The lake has modern industrial value, too; its waters cover a thick layer of diatomite, an excellent filtering and polishing agent. Diatomite is composed of dead diatoms, free-floating algae that build shells out of silicates washed into the lake from hot springs.

Icelanders dredge diatomaceous earth from the lake and pump it to a shore factory for drying. That requires heat, and Earth supplies it from geothermal steam wells almost two kilometers deep.

This geothermal zone is closely linked to Krafla's magma channels. How closely became apparent in September 1977, when molten rock shot from one of the diatomite plant's steam wells—the world's first eruption from an artificial hole. In 20 minutes three tons of lava poured out through a small opening melted in the wellhead piping. Worse, fissures opened up under the holding tanks, and workers watched helplessly as much of the winter production of diatomite disappeared into the ground.

This was the most unusual of the eruptions that had begun in the area in 1975. Until then, there had been no volcanism here since the lava flows called the "Mývatn Fires" of 1724-29. But every few months since 1975, fissures north of Krafla had been erupting for a few hours or days at a time. Scientists were excited. This was the first chance to use modern research equipment on volcanism related to Mid-Ocean Ridge spreading. Other Icelanders, however, worried; these new "Krafla Fires" threatened the lakeside town of Reykjahlíd and a controversial new geothermal power plant.

The lessons of life on an active spreading zone

The Krafla plant had opened in 1975 as an alternative to a hydroelectric dam that would have drowned a desirable valley. Those who disputed the plant's site on the flanks of Krafla went unheeded; after all, nothing had happened there since 1729.

Geothermal plants are not new to Iceland. Low-temperature—less than 200°C—facilities have heated radiators in Reykjavík homes since 1928. The Krafla plant, however, was to tap a steam field, a zone of high-temperature water. This was more complicated. Political pressures forced construction to begin before test-drilling results were in. When the first wells did reach water, it was far hotter than expected—up to 350°C—and full of machinery-eating corrosive chemicals.

Then the eruptions began. Scientists don't think drilling triggered the upheavals, but they played havoc with the steam field. Many new wells found no usable steam at all. By 1982 the plant was putting out only 15 of its planned 60 megawatts.

When Haraldur and I visited Krafla, we also met Hjörtur Tryggvason and saw his tiny laboratory next to the plant. Hjörtur is a kind of foot soldier of science, a former teacher now working as a field recorder for Iceland's National Energy Authority. We followed as he made his daily rounds of the riftmeters, seismometers, and tiltmeters scattered throughout the area.

A graph on the wall of his laboratory told the story the instruments revealed. For weeks or months before an eruption, the ground—and the graph line—would slowly rise, as liquid magma inflated the chamber beneath us. An abrupt drop meant the ground had fallen as the magma chamber deflated, and lava would be on its way to the surface.

As Haraldur explains it, "Inflation continues to such a point that pressure in the reservoir exceeds the strength of the rock. The walls of the reservoir fracture, and magma then squirts out into the fissure zone north or south of the volcano. We see only a very tiny fraction of the magma at the surface. A large proportion is injected into the fissures within the crust. That magma is what keeps the crust growing."

When we arrived, the magma chamber had inflated beyond any previous level, so we decided to stay to see what would happen.

As I sat at dinner in a Reykjahlíd hotel, I thought of the Mývatn Fires two and a half centuries ago, when this end of the lake bed rose until the water was too shallow to float a boat, and a flow from Krafla overran much of the town.

Now the shoreline nearby had risen again, stranding a rowboat among the rocks. One of the swimming holes near Reykjahlíd was too hot to use. Fissures had ripped

Saga of a Town
That Wouldn't Die

*A backyard volcano builds itself
into being on the outskirts of
Vestmannaeyjar, port and only
town on the island of Heimaey.
Within 24 hours of the fissure's
opening on January 23, 1973, ships
and planes evacuated most of the
island's 5,300 people. Volunteers
stayed to begin a six-month struggle
with the new volcano, Eldfell.*

OVERLEAF: *The steam of thermal
battle rises as seas cool Eldfell's
advancing lava. On land the thick,
blocky lava slowly bulldozed a
third of the town into ruin, but
islanders most feared this seaward
flow, which could have closed the
harbor and destroyed the town's
economy. They trained fire hoses
on the lava front, hoping to cool it,
solidify it, and thus dam the flow
behind. The tactic seems to have
worked; flows stopped short of the
harbor entrance, but not before
they had added some 2.5 sq km of
new territory to the island. The hot
new shorelines warmed normally
frigid North Atlantic waters to
bathing temperature, to the delight
of grimy cleanup crews.*

2ND OVERLEAF: *Odors of volcanic
gas and baked building materials
permeate houses buried in Eldfell's
cinders. Bulldozers and shovel-
wielding volunteers eventually dug
them out, a process that took years.*

151

Where no mountain stood before, Eldfell now looms over a reborn Vestmannaeyjar. Typical of the volcano-wise Icelanders, residents have run water pipes through the new lava—whose interior may stay hot for decades—to supply central heating for the town. And local teenagers are said to find Eldfell's warm crater an ideal trysting place.

156 through farms near the north coast, and land there dropped enough for the ocean to form a new lagoon.

That evening, Haraldur and I drove past the diatomite plant and the deafening roar of its steam wells. Up the road was a field where farmers take advantage of the heated ground to grow potatoes late into the season. In 1977, Haraldur told me, the ground grew so hot the potatoes came up half baked.

Our week of volcano-watching ended without an eruption, just more swelling below ground. Four months passed before alarm bells rang as the chamber began its characteristic rapid deflation. Seventy minutes later lava erupted from a fissure eight kilometers long.

Reykjahlíd residents, some with suitcases packed for a quick evacuation, could relax; the flow was again safely north of town.

If Icelanders had a choice, I wondered, would they vote to keep their beneficent earth-fire—and its dangers? Probably so. Even after the 1973 eruption on the south-coast island of Heimaey made headlines by burying much of the port city in ash, residents not only excavated most of their houses, they wrung satisfaction from the volcano by piping water through its smoldering lava to heat the town.

Both Krafla and Lakagígar challenge the once-assumed idea of gradual and evenly distributed mid-ocean spreading. "It turns out that the volcanic zones in Iceland are not undergoing steady rifting," says Haraldur, "they are being pulled apart in jerks."

Nor is Iceland's volcanic zone a simple, uniform rift; it is a string of central volcanoes separated by irregular swarms of fissures. As the plates pull the land apart, magma chambers under volcanoes such as Krafla and Grímsvötn, instead of erupting upward, discharge to the side through the newly opened underground cracks. Thus a central volcano can feed magma to fissures many kilometers away.

Before leaving Iceland, I wanted to visit Surtsey, the island that rose from the sea south of Heimaey in 1963. I took a boat out past the rugged cliffs and fresh lava shoreline of Heimaey's harbor. Hundreds of seabirds circled overhead. Over on the Icelandic mainland I could see Hekla's cone, still steaming from its 1980 eruption, and I thought of the medieval sailors who wondered if it was indeed the gate to hell.

From Surtsey's windworn summit, I gazed over the North Atlantic, imagining the Mid-Ocean Ridge running below these waters—south through the Atlantic, around Africa to the Indian Ocean, past Australia, and across the Pacific—the Midgard Serpent, girdling the world.

Could it be that Iceland was a model-in-miniature for the whole Atlantic portion of the Ridge? Geochemistry shows that many high points along the Mid-Atlantic Ridge—Iceland, the Azores, Ascension, Tristan da Cunha, and Bouvet—are deeply rooted volcanic hotspots. Just as Iceland's central volcanoes feed magma sideways into the rift, could these hotspots be feeding magma sideways into the spreading Ridge? Had they opened the Atlantic the way a row of spikes can split a wooden plank along the grain?

We don't know yet—and we may not find out until we understand just what a hotspot really is.

Hotspots

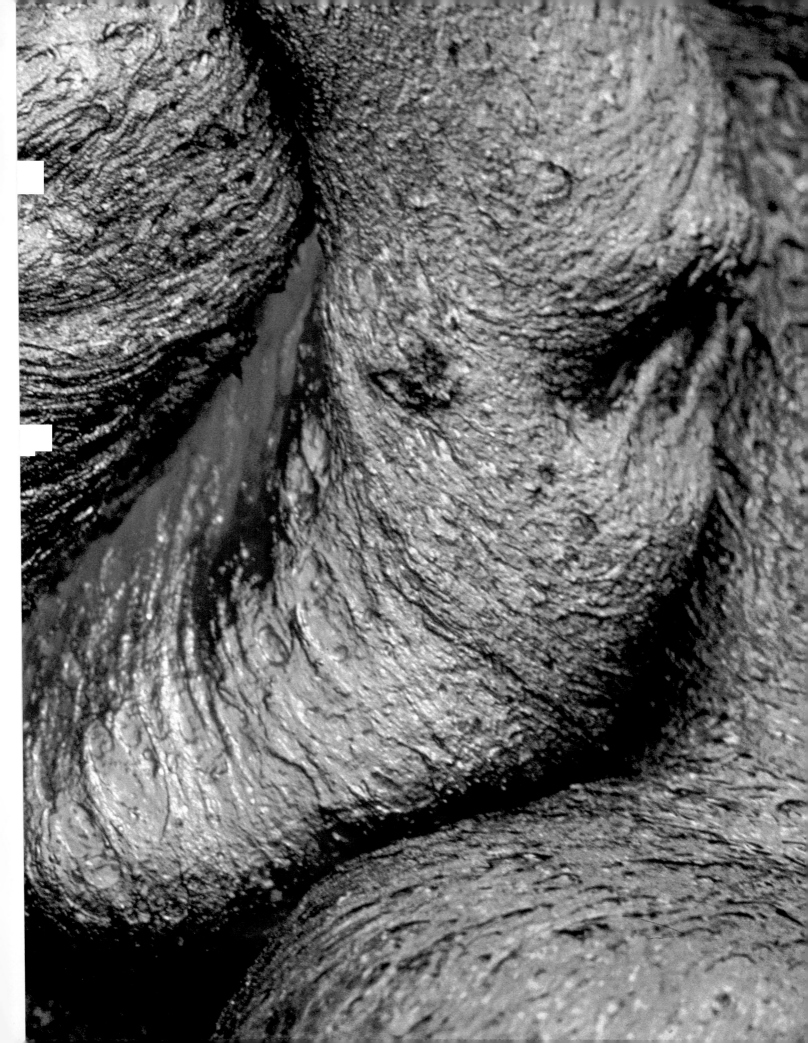

Earth's Fountains of Heat

164 **Arrived at 3 p.m. at crater's edge. Pushed across the lunaresque landscape as quickly as possible. Remoteness and danger . . . warm lava flows. . . . My shoes melted! God, I'm glad it's over!**
Field notes of photographer Steve Raymer, Réunion, 1981

Shoes often melt during a close encounter with Piton de la Fournaise. Fittingly so, for beneath "Furnace Peak" on the French-ruled island of Réunion in the Indian Ocean, a fount of magma called a hotspot burns through the underside of the Somali Plate like a welder's torch, creating at the surface a volcanic vent.

There is no subduction zone near Réunion, no seam where plates pull apart. At such places one expects volcanoes—but not here on a solid slab of crust. Hotspots explain why volcanoes arise here anyway. But scientists still debate the causes of hotspots, and even exactly what a hotspot is.

Piton de la Fournaise is one of the world's most active volcanoes, often erupting several times in a year. Well-behaved as volcanoes go, this one allows a fairly close approach by observers and tourists. There's even a trail marked with white paint to lead them to a safe lookout near the crater.

Scientists need to move in closer than that, for lava changes quickly once it leaves the vent. A few years ago, my friend French geologist Jean-Louis Cheminée pulled on a Buck Rogers suit of aluminized asbestos (complete with gold-coated glass faceplate) and sidled alone to the hellish bank of a river of molten rock. Choking gases swirled about

his head; ground temperatures as high as 500°C melted *his* shoes too. He could stay only a few minutes, but that was enough to dip out a gob of lava with a hooked rod and pop it into a container.

Such samples, taken daily, monitor the lava's evolution as the eruption progresses. Its mineral and chemical composition gives scientists a clue to the workings of the hotspot. For example, a lava of simple makeup—without the contamination of silica and other elements picked up from Earth's crust—suggests that a direct conduit from far below pumps magma to the surface before it has time to alter.

Today a small observatory keeps a constant watch on Piton de la Fournaise. Founded in 1980, it is staffed by several scientists and technicians—and by several more from France whenever the volcano stages a major eruption.

An older and more famous volcanology laboratory operates half a world away. Perched right on the lip of a crater three kilometers across, the Hawaiian Volcano Observatory has kept a constant watch on Kilauea—another of the world's most active volcanoes—since 1912. Tiltmeters, seismographs, and several generations of scientists of the U. S. Geological Survey have made this volcano on the east flank of Mauna Loa probably the most studied in the world.

The battery of instruments monitors the workings of a hotspot under the Pacific Plate. But the ancient Hawaiians—and a few islanders even today—attributed the activity to Pele, goddess of volcanoes, with a temper hot as lava.

Pele came to Hawaii about eight

centuries ago, legends say. She traveled down the chain (see map, page 166) from Kauai, northwesternmost of the inhabited isles, to the "Big Island" of Hawaii at the chain's southeastern end. Although science dates these islands in millions, not hundreds, of years, fact and legend strike an uncanny parallel. Both recognize that the islands change character along Pele's path.

Travels of a goddess mirror Hawaii's growth
Journeying southeast from Kauai after a family squabble among the gods, myths say, the fiery Pele landed at the next island in her path, Oahu, and there dug a fire pit with her magic *paoa,* or spade. But her spiteful sister the water goddess drenched the fire. Down the chain of islands Pele wandered in search of a home, digging herself a pit on each one in turn.

Finally she landed at Puna, today a sprawling seaside district of cane fields and lava flows on the Big Island's eastern flank. Upslope she climbed to visit the Wood Eater, god of the island. But he had fled in fear of Pele's wrath—and was never seen again.

Pele began to dig. After many days and nights of toil she excavated within Kilauea's caldera a great fire pit, Halemaumau, "beneath whose molten flood, in halls of burning adamant and grottoes of fire, she consumed the offerings of her worshippers. . . ." And there she has dwelt ever since.

Now compare legend to science. In 1849, American geologist James Dana began to document an age progression in the islands that fit the legends of Pele's travels. He

noted that, as an observer moves
from Kauai to the Big Island, the
volcanoes appear less eroded and
therefore younger.

Ian McDougall, an Australian
scientist, later dated the rocks on
the Hawaiian Islands and found that
Kauai is at least 5.6 million years
old. But follow Pele southeastward
and you find, as McDougall did,
that the islands become younger.

The dormant volcano Haleakala,
the "house of the sun," soars 3,056
meters (10,026 ft) above the shores
of Maui, opening into a crater 32 kilometers around—so big that sight-
seeing airplanes once flew down
into it. Maui is less than two million
years old; Haleakala last erupted in
the late 1700s. And the Big Island,
Maui's next-door neighbor, is less
than three-quarters of a million
years old; it still layers itself in lava
from Mauna Loa and Kilauea.

These two are the only Hawaiian
volcanoes above water to erupt in
this century; they now lie over the
Hawaiian hotspot. But far earlier
eruptions have left a much longer
trail than the one Pele traveled.

Northwestward from the verdant
slopes and valleys of Kauai stretches the rest of the Hawaiian chain, a
dotted line of stark pinnacles, sand
islets, and coral atolls drawn across
some 1,900 kilometers (1,200 mi) of
ocean. As islands, these virtually
uninhabited Leewards—named of-
ficially the Northwestern Hawaiian
Islands—are not very impressive.
But as mountains they rank among
the world's biggest. Like their inhabited younger siblings, they rise
from a seafloor six kilometers
down. From end to end, the entire
Hawaiian chain includes more than
50 volcanoes both above and below

166

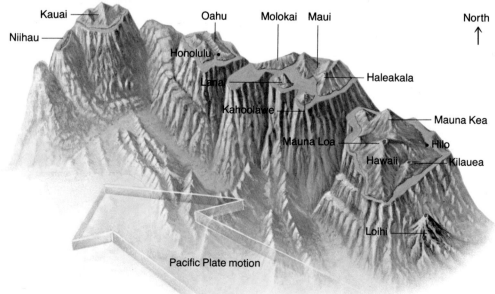

the surface of the Pacific Ocean.

Two things about this long mountain chain troubled early plate tectonics theorists. First, the ocean floor on which it stands is a lot older than the chain itself, 80 million years older near the Big Island and up to 95 million years older beyond the Midway Islands. And second, the Big Island's active volcanoes do not stand where the theory of plate tectonics said they should: at the edges of the plates—either where the plates are spreading, as along the Mid-Ocean Ridge, or where they are colliding, as in Japan. Instead, Hawaii stages its pyrotechnic displays squarely in the middle of the Pacific Plate.

Canadian scientist J. Tuzo Wilson, one of the pioneers of plate tectonics theory, suggested in the 1960s that convection in the underlying mantle produced what another scientist later dubbed a "hotspot," and that the islands arose from magma poking through the seafloor as it moved over the hotspot. This would explain why the chain increases in age away from the Big Island's active volcanoes, and why the seafloor beneath is so much older: That seafloor had formed millions of years earlier and thousands of kilometers farther east, at the East Pacific Rise.

A few years later, W. Jason Morgan of Princeton University added further ideas to a scientific controversy that continues today. Morgan postulated a thunderhead of concentrated excess energy beneath the Big Island, welling up from deep within the planet. This thermal anomaly created a plume of heated rock that cut through the plate like a blowtorch, building up one volcano

A fragile tendril of sand, tiny Whale-Skate Island lies atop French Frigate Shoals in the Northwestern Hawaiian Islands. Little but coral remains of this volcanic island worn flat by the sea during its 12-million-year odyssey from the hotspot, now a thousand kilometers away.

after another as the moving plate passed over the plume and bore the mountains away like newly baked cupcakes on a conveyor belt.

Morgan went further, arguing that the Hawaiian archipelago was only the more recent part of the hotspot trail; that the torch under the Big Island had even earlier blazed a trail of 30 more volcanoes, now worn away to undersea mountains: the 2,500-kilometer (1,600-mi) Emperor Seamounts chain.

The oldest of the Hawaiian seamounts is about 37 million years old. The youngest Emperor seamount is just 4 or 5 million years older—a neat chronological fit. And like the Hawaiian chain, the Emperor Seamounts seem to grow older steadily. The ones farthest north, where the Pacific Plate disappears into the Aleutian Trench, are about 70 million years old.

Adding the Emperor line to the Hawaiians bothered some scientists. The Hawaiian-Emperor track has a bend in it, a big one. But to Jason Morgan the bend wasn't a problem; it was telling us something. He hypothesized that the hotspot is virtually stationary. Thus the bend tells us that the Pacific Plate changed direction at about the age of the mountains at the bend—in this case about 40 million years ago. Before then, the plate, which is now creeping northwest, had been moving almost due north.

Other research has since supported Morgan's basic contention: Hotspots stay in roughly constant positions relative to Earth's mantle.

Of all the hotspot islands, we know the most about Pele's home, Hawaii. There in Kilauea's huge caldera yawns the fire pit dug by

Earth's Tallest Mountain

Skiers slide down Hawaii's roof on *snows that cap Mauna Kea for a few winter months. Earth's tallest mountain, Mauna Kea rises 10,203 m (33,476 ft) from seafloor to summit. Near neighbor to active Mauna Loa, this volcano has lain dormant for 3,600 years. The skiers traverse a cinder cone, one of 132 that stud the flanks and summit of the shield volcano, built layer by layer from countless lava flows.*

OVERLEAF: Lava from the depths of Hawaii spews onto the floor of Kilauea's caldera in the spring of 1982. Hurtling blobs of incandescent lava sometimes rain down on the area around a fire fountain. Despite such dangers, scientists and tourists flock to Hawaiian eruptions—showy outpourings that rarely explode with deadly violence.

2ND OVERLEAF: Ropes of stone, typical of pahoehoe (pa-HOY-hoy) lava, coil near Kilauea's Mauna Ulu crater. The surface of this flow forms a skin as it cools, then wrinkles as the lava creeps on. Pahoehoe can turn to a type called aa (AH-ah) as cooling hardens it and churning breaks the lava into sharp chunks—a one-way process that rarely turns aa to pahoehoe.

Pele, a great well half a kilometer deep and nearly twice as wide. Her "grottoes of fire" we identify as a large magma chamber, a long, shallow reservoir three to six kilometers beneath the caldera's floor. And now we know what lies even farther down.

Tilts and quakes
foretell a fiery show

As in Iceland, the chamber slowly balloons as magma flows into it. Tiltmeters, which act like giant carpenter's levels, measure swelling that can distort the surface as much as a meter and a half. During the eruption, the reservoir deflates as magma flows toward the surface. It may take days or even years after the eruption to recharge the reservoir with magma. Then the telltale pattern will repeat: slow inflation, rapid deflation.

When magma rises to refill the chamber, it pushes against the hard rock of the volcano's inner plumbing, cracking the walls and generating micro-earthquakes. The waggling needles of strategically placed seismographs record the tremors, helping observatory scientists pinpoint where the cracking occurs.

From more than 30,000 micro-earthquake recordings, they have assembled the best three-dimensional image ever made of a volcano's innards, a shadowy outline of Pele's grottoes of fire. From a deep magma zone at least 60 kilometers down, a vertical conduit several kilometers in diameter climbs toward the surface, twisted and kinked like a badly worn crankshaft. About seven kilometers below the surface, this primary conduit divides into two channels and opens into the

shallow magma chamber. Ducts lead from there to the erupting vents at the surface: Halemaumau, Mauna Ulu, and other vents on Kilauea's rift zones, including the pits along Chain of Craters Road in Hawaii Volcanoes National Park.

I drove that road in 1982, stopping amid a surrealistic landscape to walk to Mauna Ulu crater on the southeast side of Kilauea. Hiking through a dreary mist over still-warm lava, I stepped quickly to keep the heat from coming through my shoes. At the crater rim I found an eerie wall of steam rising from fissures around the edge. I couldn't see through this curtain, so I picked up a rock and threw it into the crater. I began to count, one thousand one, one thousand two . . . one thousand five—*plunk!* Five seconds, or about 135 meters (440 ft) to the crater's bottom. Suddenly I realized I was alone, miles from my car and a single step from the edge of the pit. If I slipped, no one would know. Quickly I moved back.

Kilauea, focal point of the park, perches on the eastern slope of another active volcano, Mauna Loa. Built flow by flow from the seafloor, Mauna Loa is the world's largest single mountain, with a volume of over 40,000 cubic kilometers. Both Mauna Loa and its taller neighbor, Mauna Kea, rise more than 10,000 meters from the ocean bottom—taller than Everest from base to tip by well over a kilometer.

From the oldest of the Emperor Seamounts to the latest lava flow from Kilauea, this great dogleg range testifies to about 70 million years of Pacific Plate history. But the plate is still transporting the whole assemblage northwestward

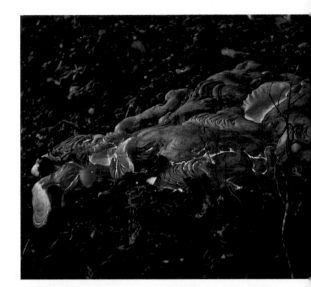

from the hotspot underneath. Will a new island one day appear? Actually, it is already in the making: Loihi Seamount, 30 kilometers (19 mi) southeast of the Big Island.

Observatory scientists watching Loihi in the 1970s recorded earthquake swarms there—swarms believed to herald a submarine lava eruption or the movement of magma within a volcano. In 1978 scientists towed a submersible camera over Loihi and brought back photographs of a young, glassy surface, a confirmation of recent lava flows.

Later they dredged the flanks of the seamount and brought up a variety of lava samples—and a surprise: a type called alkalic which was thought to erupt in Hawaii only during a volcano's dying stage. Alkalic lava is rich in volatile gases, which might well help the magma punch through the lithosphere and rise to the ocean floor. Thus a Hawaiian volcano may begin with the same kind of lava that fills in its collapsed caldera as it matures and dies.

"As beachfront properties go, the warm, submarine flanks of Loihi . . . are a terrible investment," warned *Science News* magazine in December 1981. Hawaii may have to wait from 2,000 to 20,000 years for its new island to grow another 950 meters (3,100 ft) and break out into the sun and air.

Tears and tantrums leave lasting mementos

The Hawaiian Islands' unique geologic features have inspired a rich folklore, and on the Big Island the ill-tempered Pele reigns at its violent center. Ancient Hawaiians found blobs of lava that had cooled in teardrop shapes as they spurted

into the air, and called them Pele's tears. Fine filaments of syrupy lava drawn from flying blobs surely must be Pele's hair. And once, when Pele lost a race, she lost her cool as well; stamping her foot, she called up an awesome lava flow that turned onlookers to pillars of stone.

Pele's petrified people still stand today in Lava Tree State Park in the Puna district. They are actually tree molds from a 1790 eruption that engulfed a forest. When the liquid lava surrounded a tree, it would cool and harden around the trunk, setting it on fire in the process. The rest of the lava would flow on, leaving tree trunks to burn or rot away. Only the stone jackets remained, eerie pillars turned to people by fearful worshipers.

So powerful a goddess demanded many sacrifices and imposed many taboos. But missionaries arriving in 1820 soon put a stop to that. Four years later the Christianized Princess Kapiolani defied Pele's taboos: Striding across the steaming floor of Kilauea caldera, she stood on the lip of Halemaumau and brandished the sacred berries that only males were permitted to touch. "Into the flame-billow," recounted the poet Tennyson, the princess "dash'd the berries, and drove the demon from Hawa-i-ee."

Not quite. In November 1880 an enormous lava flow poured out of Mauna Loa and advanced downslope toward the city of Hilo. Alarmed citizens coaxed Princess Keelikolani down from Honolulu to sing the ancient chants to Pele and pour a bottle of brandy on the flow. The next morning the lava halted short of the much-relieved city. When Mauna Loa erupted in 1935

178 and again threatened Hilo, a more modern approach turned the fiery tide. Observatory director Dr. Thomas A. Jaggar attacked the advancing lava with ten U. S. Army Air Corps bombers, each carrying two 600-pound bombs. The first attack blasted the flow near the upper vent; the second sortie struck farther downstream. The disrupted flows poured into new channels that spread harmlessly over the mountain's barren upper slopes.

Berries, brandy, or bombs notwithstanding, there are still those who say they have glimpsed Pele just before an eruption.

Volcanology in scuba gear

In 1973, James G. Moore of the U. S. Geological Survey took me on an unusual tour of Kilauea. First we walked across the crater floor to the edge of Halemaumau. The fire pit had been almost continuously active for a century after the first visit by Europeans in 1823, but since then it has become more intermittent. Even so, steam rose and the smell of sulfur filled the air. The sight was impressive, but my main reason for coming to the island was to see Kilauea in another way: from beneath the sea.

Just the year before, Mauna Ulu had erupted, but unlike most recent flows, this one scorched down the slopes all the way into the ocean. Jim and photographer Lee Tepley had pulled on scuba gear for an attempt to make the first film ever of lava fields forming underwater.

Jim took me by boat to where the flows had crossed the shoreline. He described their earlier experience as we prepared to dive. At the surface, he said, the water had been

hot. It was tinged dark yellow near the lava and tasted sickeningly of sulfur. About six meters (20 ft) underwater, they found the lava front advancing down a coral-encrusted slope. It looked, Jim recalled, like black toothpaste being squeezed from a tube. The tube would crack open at the end for just a few seconds as red lava poured out and instantly cooled, building in turn a new section of tubing—the characteristic pillow lava.

Suddenly the water pressure collapsed a tube in an implosion that sent out a shock wave like the blast of a grenade. Jim floated limply for a moment, stunned by the concussion. Soon he recovered and rejoined Tepley to get their pictures.

Now he was about to return, this time to a flow long since frozen into hardened stone. My dive with him was much less eventful than his earlier one, but it gave me a better understanding of how red-hot lava, cooling instantly when it hits cold seawater, forms the glassy pillows of basalt I have since seen on my own undersea explorations.

On the island of Hawaii, superlatives accrue like lava flows. During a 1959 eruption just east of Kilauea's caldera, a gigantic fountain of incandescent lava spurted up to 580 meters (1,900 ft). One massive outpouring from Mauna Loa laid down enough lava to pave a two-lane road nine times around the globe.

Flows that boil and hiss into the sea may be shattered by steam explosions or worn by storm and wave into bits of volcanic glass that later return to land as one of the famous black sand beaches. A few stretches of shoreline turn green with crystals of olivine—a delight

Life Abounds in a Hotspot Home

Time stands still for tortoises drinking rainwater from a pool in the Galápagos Islands. Fumaroles in the distance recall a violent volcanic past here in the Alcedo caldera on Isabela Island. A hotspot may lie almost directly below: Alcedo and four other volcanoes that made this island have all erupted in the last 200 years. Like some other hotspot chains in the Pacific, these islands are spaced about 35 km apart.

OVERLEAF: *"A hideous-looking creature, of a dirty black colour," wrote Charles Darwin of the marine iguana of the Galápagos Islands. Its beauty lies in its swimming and diving skill. It is the only lizard that eats seaweed. Although iguanas depend on sunlight for warmth, midday equatorial heat overcomes them. To compensate, they seek out clefts washed by incoming waves or shaded by overhanging ledges of hardened lava. When even those refuges prove inadequate, they swim and feed until, numbed by cold water, they return to land for another session of sun.*

Cratered by the vintner's spade, lava cinders shield grapevines from whipping winds on Lanzarote in the Canary Islands. The loose granules sop up precious rain and dew, holding it for dry spells. Low walls of rocks add protection from the wind. A hotspot under the Canaries last revealed its power in volcanic eruptions 150 years ago.

184 to tourist and photographer alike.

And so are the spectacular eruptions that spawn them. People keep up with volcanic activity here; they can even phone Dial-a-Volcano for hot news of the latest tremor. They run *to* an eruption rather than away, for only rarely does the water in a Hawaiian volcano's plumbing system build up enough steam pressure to cause an explosive blowout.

That happened in 1790 as a platoon of Hawaiian warriors marched across the Kau Desert just southwest of Kilauea caldera. Heat and dust clogged their lungs and steam condensed to rain as they took their last halting steps through the thickening ash. About 80 perished—but their poignant footprints remain, baked hard by the desert sun. Some of the prints are now sheltered under a pavilion at Hawaii Volcanoes National Park.

Nowhere else on earth does the hotspot theory seem quite so well documented as in the Hawaiian island chain. Yet the planet appears to be peppered with hotspots. Tuzo Wilson, using the most basic definition of a hotspot—volcanism far from plate edges—listed just a few, all in the oceans. Jason Morgan expanded the list to about 20, looking around the world for places where the force of gravity was unexpectedly strong—the locations, he said, of thermal plumes formed by convection cells in the mantle.

Some scientists now list more than 120 hotspots. Some of these they identify by the earlier earmarks of a hotspot. Others they infer by the telltale presence of broad domes several hundred kilometers wide pushed up by the rising magma. Their lists include some likely hotspots under Africa and Europe; in North America, under Yellowstone Park; and in the oceans, under the Azores and the Society and Galápagos Islands.

A hotspot under the Galápagos? That takes some explaining, for these islands, a few hundred kilometers east of where we discovered the deep oases of the Mid-Ocean Ridge, don't line up in a trim file; they seem scattered almost like sown seeds.

Many experts believe the apparently chaotic Galápagos actually tell of an orderly process. The Galápagos ridge, they point out, doesn't stay in one place; it keeps shifting farther north, a few kilometers at a time. Thus, some islands would have built up over the hotspot on the edge of one plate and been borne away northward. Then, when the ridge jumped to a place just above the hotspot, islands arose right on the seam and moved away from it in both directions. Finally, the seam jumped again, and the newest islands were built up on the southern-moving plate.

Hotspots may be cogs in the ponderous engine that drives the moving plates. For as magma from the lower mantle wells up toward the surface, much of it never gets there. Instead it fans out under the plate, according to Jason Morgan's hypothesis, thus buoying it and helping to push it past the hotspot.

An active volcano, by one definition, is one that has erupted within the last 10,000 years. Some 600 of them pock the face of Earth. But only about 60 erupt in an average year. Piton de la Fournaise usually makes the active roster; Yellowstone never does. And yet. . . .

Plumes of water and steam born in Yellowstone's volcanic basement shoot from Castle Geyser. An underlying hotspot powers the park's geysers and hot springs, heating water that percolates several kilometers below the surface. Driven by expanding steam, seething water bursts upward through a maze of cracks.

As pressure drops near the surface, the superheated water flashes into steam and erupts in a tumultuous display. Geysers only hint at the volcanic might lurking beneath the park. Huge explosions have ravaged the region three times in the last two million years. Scientists think another may be due—perhaps soon, geologically.

186 **Yellowstone: time bomb in the Rockies**

And yet geologists consider Yellowstone a powder keg, perhaps ready to explode and join the list.

Indeed, it is more than a powder keg. Steam and hot water working their way upward from Yellowstone's depths bring tangible signs of a volcanic force as great as any recorded on earth. Major eruptions long ago wracked the Yellowstone region; the last spread a layer of ash as thick as a down comforter hundreds of kilometers away. Another eruption may be due, many geologists say. They look at Yellowstone today and wonder about tomorrow.

John Colter looked at Yellowstone and wondered, too. Probably the first non-Indian to visit Yellowstone, he ventured into the region in 1807 and found a series of valleys with more than 10,000 hot pools, geysers, mudpots, and gas-spitting vents. For the next half century, other visitors brought back stories of incredible sights to skeptical Easterners. "The whole country beyond," reported trapper Joseph Meek after wandering through in 1829, "was smoking with vapor from boiling springs, and burning with gases issuing from small craters, each of which was emitting a sharp, whistling sound. . . . Out of these craters, issued blue flames and molten brimstone."

Moved by photographs from an 1871 scientific expedition, Congress wisely voted to preserve Yellowstone's natural beauty in the world's first national park. Today it stands unparalleled among world geyser areas. Nearly two-thirds of the world's active geysers spew from Yellowstone's vents, including the largest, the most frequent, and the most powerful.

A devilish spirit seems to infuse the displays. "In walking among and around them," wrote an early explorer, "one feels that in a moment he may break through and be lost in a species of hell."

Years of research unmasked the demons, revealing instead the geologic forces under Yellowstone's simmering surface. Three titanic explosions have rocked the region, on a fairly regular schedule of once every 600,000 to 800,000 years. The first and largest of them occurred about two million years ago. Within days, 2,500 cubic kilometers (600 cu mi) of dust, ash, and lava blew skyward—an estimated 15,000 times the amount blown out by Mount St. Helens in 1980. Magma exploded again 1.2 million years ago. And again 600,000 years ago. Numerous smaller eruptions as recently as 70,000 years ago have since covered remnants of the earlier ones.

Geologists ponder Yellowstone's repeated destruction and forecast a gloomy future. To learn more about a probable next blast, they look at what the previous ones left.

A crater too big to be noticed

Not until fieldwork in the 1950s and 1960s did geologists discover an amazing feature: Most of central Yellowstone Park lies inside a volcanic caldera left by the last major eruption, an immense collapsed crater reaching far beyond individual volcanic vents. They made the find after years of study, even making tedious comparisons of one rock to another. More detailed field studies a decade later confirmed the discovery. Dense forests and enormous lava flows obscure the caldera's edges—it is simply too big to be seen from ground level.

Scientists have also discovered remnants of other calderas left by the earlier eruptions. Huge cavities opened beneath the surface as each explosion blew away the underlying magma. The ground fell into these gaping caverns to create the calderas. The last explosion left an elliptical caldera 45 kilometers wide and 75 kilometers long (28 by 47 mi).

Wielding their seismic sensors, geophysicists explore the earth beneath the calderas. When X rays pass through a human body, bones absorb some of them, leaving a shadowy image on the recording film. When seismic waves—such as those generated by an earthquake—pass through the earth, molten rock slows the waves. Using a series of seismographs around the park, the scientists can record the delays to put together a shadowy outline, a kind of seismic X ray, of heated rocks beneath the park.

By the early 1970s, after 15 years of studying such X rays, geologists had a general picture of the earth beneath Yellowstone. Two magma chambers lie side by side under the caldera. Together they hold enough magma and hot rock to bury all Wyoming some 35 stories deep. Beneath the chambers, a column of heated rock plunges 200 kilometers (125 mi) into Earth's top two zones —through the lithosphere and into the asthenosphere.

Scientists had trouble putting together a clear picture of the column below 200 kilometers. Then, in October 1975, help came from an

Carried away from a mid-ocean ridge on newborn seafloor, two lines of seamounts recorded early eruptions of what some experts speculate was the Yellowstone hotspot (upper left). As the North American Plate overrode the ocean floor, it planed away the seamounts (upper right). Today, their basaltic remains lie amid the

Pacific Northwest's Coast Range. An erratic trail of volcanic debris may trace the continent's later motion over the hotspot (lower). Yellowstone's last big caldera eruption, 600,000 years ago, left further evidence of its power—a thousand cubic kilometers of ash strewn across the shaded area.

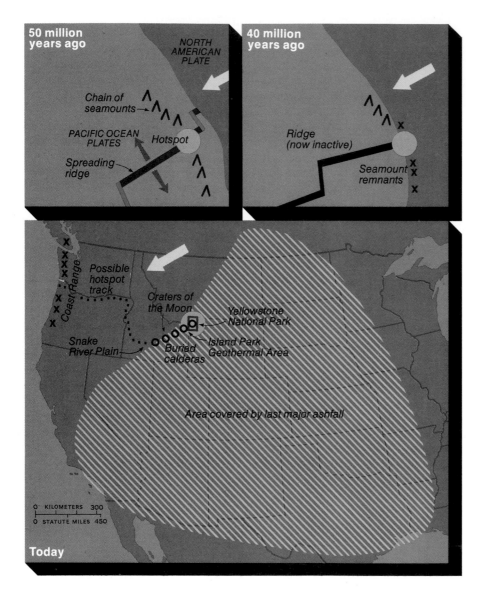

unusual quarter: a nuclear bomb test in Nevada. Instruments were readied near Regina, Saskatchewan. Because of Earth's curvature, that seismic vantage point meant that shock waves speeding in a straight line from blast to sensors would pass right under the known column at the midpoint of the journey. Gauging the bomb's mighty jolt much as they had gauged the shallower tremors before, the scientists found that the cylinder underneath Yellowstone stretched at least another 200 kilometers down.

New answers, new questions

Varied evidence had first suggested this hotspot idea. Traces of volcanic rock dotting Idaho—and perhaps reaching into Oregon and Washington—lined up in a track like the one drawn by the Hawaiian-Emperor line of islands and seamounts. The farther from Yellowstone you go, the older are the earliest volcanic rocks in each area.

The nuclear blast study and other seismic research gave direct evidence of a column of heated rock stretching into the mantle. In Hawaii, geologists had modeled the upper 60 kilometers of a volcano's inner workings from island vantage points. Yellowstone's location on a continent made it easier for scientists to stand back far enough to see what lay hundreds of kilometers farther down in a volcano's roots.

It also gave them a better understanding of continental hotspots in general. At least one geologist has suggested that continents act like insulating blankets, trapping heat in the asthenosphere below until temperatures rise enough for a hotspot to form and sear right through the

Warm water turns a Yellowstone creek into a natural hot tub for a park worker bone weary from winter chores. Here the scalding water of thermal pools has mixed with the cold stream water; in such places swimming is sometimes allowed. But strict rules forbid even dabbling a toe in most thermal waters. In these the mineral formations can be as fragile as they are beautiful, and the searing heat can kill—two good reasons to enjoy the park's thermal features at a safe distance.

continental crust. Thus the continents may be sowing the seeds of their own destruction.

The most popular theory of hotspot origin was still Jason Morgan's deep-mantle convection. But some said the nuclear blast results didn't fit his theory; the data suggested instead that the lower half of the column held radioactive elements such as potassium, thorium, and uranium, leftovers from the sorting of material when the planet was taking shape. So a new theory proposed that it is the radioactive decaying of this "chemical plume" that heats the rocks closer to the surface.

Geologists by now agreed that a hotspot lay beneath Yellowstone, but many disagreed with the new explanation—and controversy runs deep even today. Many theorists continue to argue for a plume anchored deep in the mantle. But some say its heat comes from chemicals, others say from thermal currents. And still others offer completely different explanations.

These scientists do not agree that a plume in the mantle exists at all, whether thermal or chemical. Some attribute the hotspot's trail of volcanic debris to continental rips; as the crust tears apart, pressure drops and rocks tens of kilometers deep begin to melt. Others argue that a rough spot on the underside of the plate generates heat as it rubs across the asthenosphere below. Small pockets of magma formed by this heat float upward. Again pressure drops, triggering more melting. This feedback leads to a self-perpetuating "thermal runaway" and—at the surface—the hotspot trace.

Whatever drives the Yellowstone hotspot, scientists agree that it

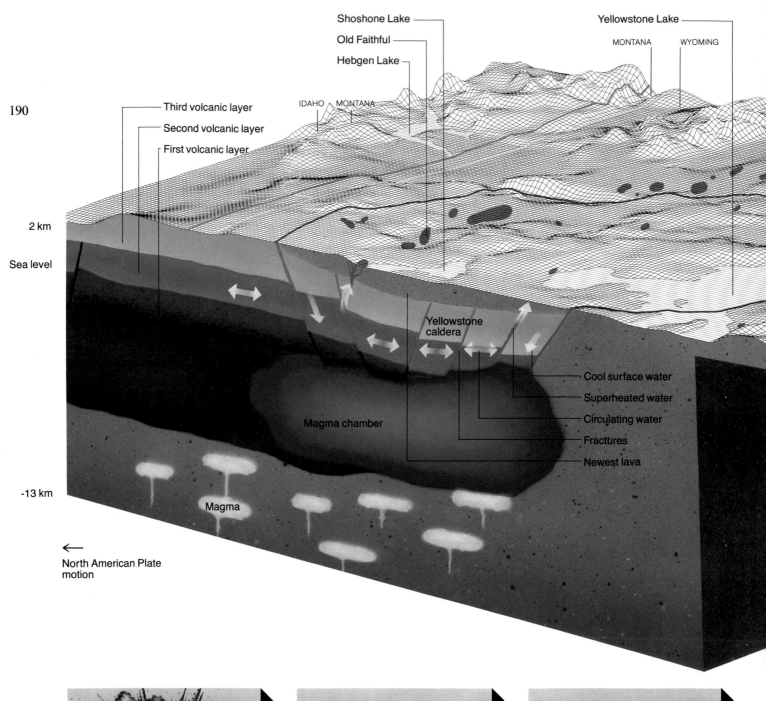

190

Third volcanic layer

Second volcanic layer

First volcanic layer

Shoshone Lake

Old Faithful

Hebgen Lake

Yellowstone Lake

MONTANA   WYOMING

IDAHO   MONTANA

2 km

Sea level

Yellowstone caldera

Cool surface water

Superheated water

Circulating water

Fractures

Newest lava

Magma chamber

-13 km

Magma

← North American Plate motion

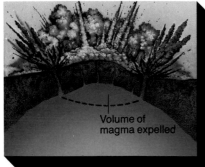

Volume of magma expelled

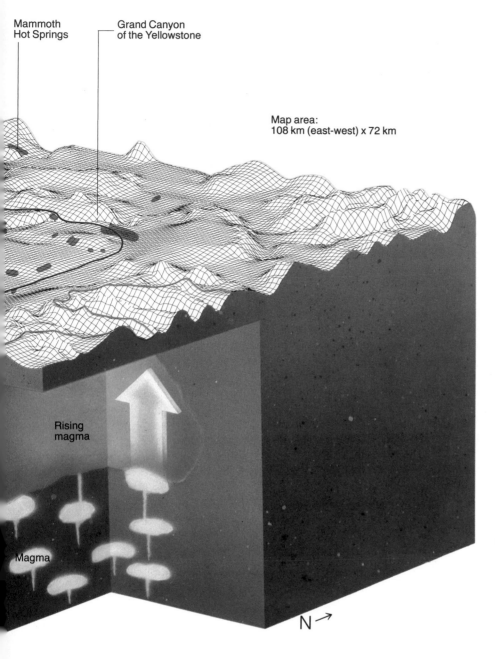

Mammoth
Hot Springs

Grand Canyon
of the Yellowstone

Map area:
108 km (east-west) x 72 km

Rising
magma

Magma

N →

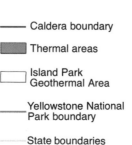

*Birth of a caldera: As swelling
magma splits its earthen cover,
concentric cracks release pent-up
gases, powering an explosive
eruption (far left). The dome of the
emptied chamber collapses,
leaving a wide caldera hundreds of
meters deep (center). Sometimes
additional lava oozes through the
faults (near left), resurfacing the
caldera or hiding it entirely.*

— Caldera boundary

▨ Thermal areas

▭ Island Park
Geothermal Area

— Yellowstone National
Park boundary

⋯ State boundaries

Yellowstone:
Pressure Cooker
With a Fragile Lid

*Rising magma bulges in a chamber
beneath eastern Yellowstone
National Park, threatening
eventual explosion. Older magma
in the same chamber slowly cools
beneath the 600,000-year-old
Yellowstone caldera. Seismic
studies revealed two of these huge
chambers of viscous rhyolitic
magma under the park. Below
them lie smaller pockets of
more fluid, less explosive basaltic
magma. For hundreds of
kilometers beneath the park, a
column of hot rock extends into the
asthenosphere. North America's
southwestward drift apparently
transports solidified magma away
from the hotspot.*

*The chamber below Yellowstone
caldera stokes the boilers beneath
Old Faithful and other geothermal
attractions. Groundwater seeps
through faults in the volcanic
layers left from three cycles of
eruptions, sinking on a years-long
journey toward the magma
chamber. In the high pressures 2 to
3 km deep, water simmers at three
times its boiling point at the
surface. Eventually heat lowers its
density and the superheated water
begins to rise.*

*The topmost 100 meters, where
the water can boil, shapes the final
display. If constricted, channels
send water and steam directly to
the surface, and geysers spout.
Along wider cracks, subterranean
pools may absorb the boiling
water, and a quiescent hot spring
gurgles above. When only a trickle
of water reaches the surface,
strong acids turn the soil into
bubbling mudpots. If all the water
boils off below, vapor escapes
from fumaroles and steam vents.*

A Curiosity Shop of Minerals and Microbes

Tubes of travertine, a form of calcium carbonate derived from limestone, stand 6 cm (2 in) tall in the mineral-laden waters of Yellowstone's Mammoth Hot Springs. Hot water dissolves subterranean limestone, spewing out two tons of minerals a day. As the water cools, the minerals in it precipitate, forming spires, terraces, and other structures.

OVERLEAF: *Bacteria and algae thrive in the boiling waters of Mammoth Hot Springs, painting travertine terraces with muted colors. Sometimes the organisms paint with brighter hues. "It was of three distinct colors. . . ," wrote trapper Osborne Russell of a Yellowstone hot spring in the mid-1800s. "For one third of the diameter it was white, in the middle it was pale red, and the remaining third . . . light sky blue." Such microscopic life feeds on sulfur and other chemicals; thus, perhaps, may life have begun in primordial seas.*

2ND OVERLEAF: *A stew of volcanic clay, sulfur gases, and water simmers in a Yellowstone mudpot. Minerals tint the mudpot white to gray; sulfur adds a dash of yellow; iron stirs in red, brown, and black. Mudpots range in consistency from nearly dry "mud volcanoes" to soupy brews such as this one.*

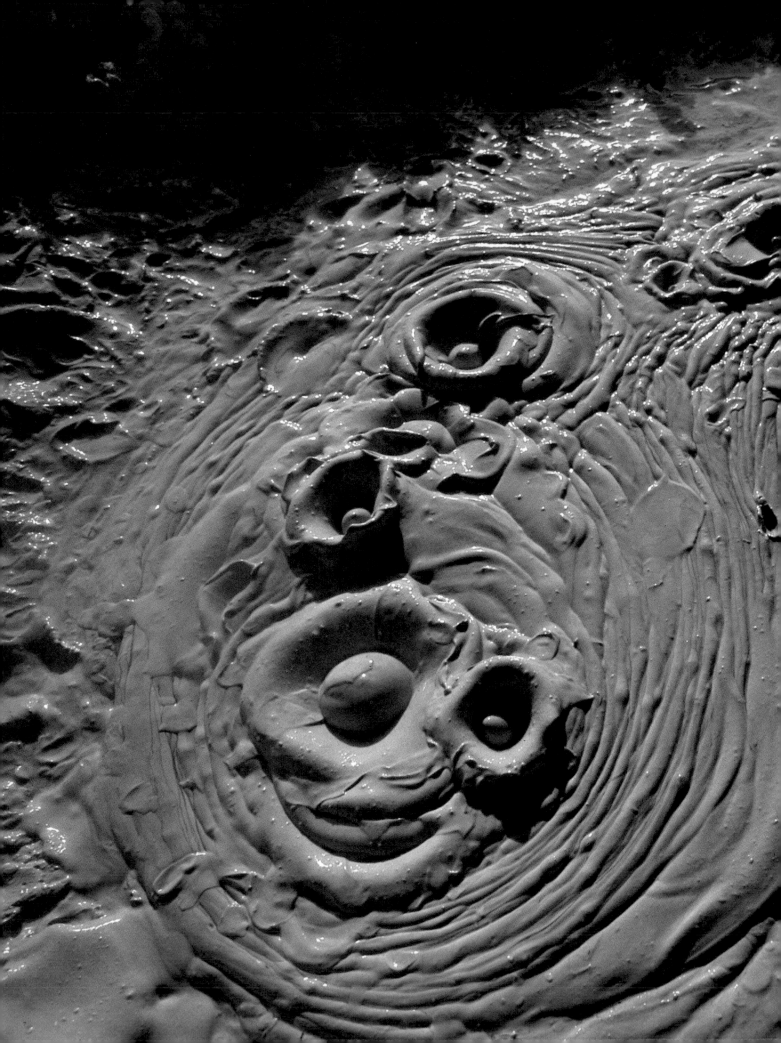

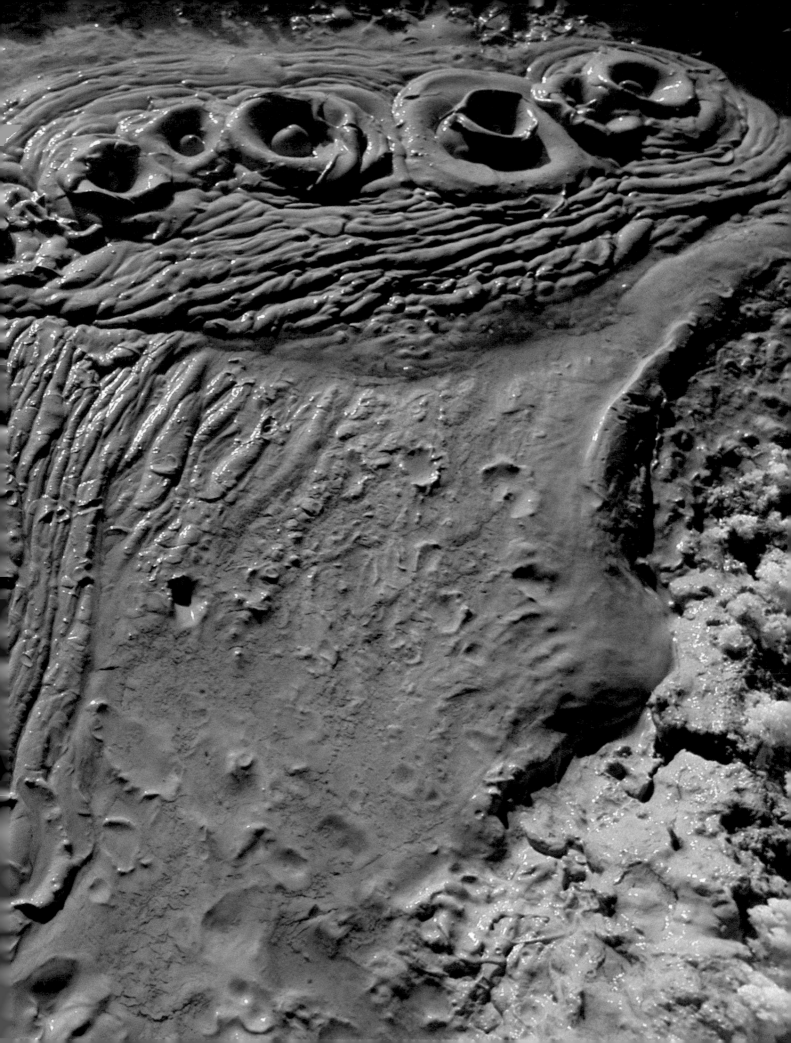

Spray from a nearby thermal vent freezes over fresh snow to shroud a dead lodgepole pine, attesting to the dangers of life near such fickle heat sources. Warm water may nourish life for a time, then destroy it with scalding steam. Heavy snows and earthquakes can alter the plumbing of a spring, changing its character overnight.

breached the surface near Boise, Idaho, about 14 million years ago. The continent overhead moved to the southwest at about 4 centimeters (1.6 in) a year, leaving a hotspot trace along what is now the eastern Snake River Plain. Lava flooded the plain again and again.

This sequence suggests to scientists an explanation for Yellowstone's long volcanic silences and short cataclysmic eruptions. They conclude that rhyolitic magma predominates in the chambers beneath the caldera, because rhyolite is less dense than basalt and thus rises higher from the mantle. Steam and carbon dioxide aerate the magma, building explosive power. Denser magma lurks deeper in the crust, adding heat to the rhyolite above. Bulging beyond capacity, the upper chambers fracture the surface. Cracks tap the pressurized gases, triggering an eruption. Only after the rhyolite hardens does the basalt flood out through new faults.

Magma left in the ground continues to heat Yellowstone, the latest point to rest over the hotspot. A network of cracks collects rain and snowmelt to feed thermal displays.

Life in the devil's home

The hellish heat below also helps maintain a wildlife paradise above. Elk, bears, birds, and fish thrive in the park; some animals linger by the heat vents when winter bares its teeth. General William Tecumseh Sherman pronounced the Yellowstone River "the best trout-fishing stream on earth."

Dense forests flourish in the mineral-laden soil. Steam wafts from geyser basins. Hot water flowing through rhyolite brings silica from the quartz-rich rocks; as the water surfaces in springs and geysers it deposits the silica in formations of whitish-gray sinter, known also as geyserite. When water circulates through limestone, it deposits calcium carbonate as travertine. Algae, bacteria, and minerals often tint the pools and formations. From "yellow stones" of rhyolite comes the tinge that gives the park its name.

Heat brings more than beauty; it also is a form of energy. Just before the continent moved the Yellowstone region over the hotspot, what is now called the Island Park Geothermal Area in Idaho lay over the rising magma. Stores of heat may remain under the area, souvenirs perhaps of its stay over the hotspot. Commercial developers in the late 1970s filed over 200 applications to try tapping that heat for energy.

But delicate tolerances govern geothermal systems. Island Park and Yellowstone are connected by faults that formed the caldera of two million years ago. Geologists ask whether hot water taken from Island Park might dry up Yellowstone's natural displays.

Many people see a clear warning in the precedents. When New Zealand opened the Wairakei geothermal plant in 1958, the nearby Great Geyser, fifth largest in the world, stopped spouting within a decade. The Beowawe Geysers of northern Nevada once ranked second only to Yellowstone on the North American continent. In the 1940s, geothermal researchers started drilling test holes there. By 1961, no geysers remained. Hot springs in Italy have met similar fates.

Even some geysers in Iceland have lost their force due to changes

202

in their geothermal environment—an irony indeed, since all geysers take their name from an erupting Icelandic spring named *Geysir*, meaning "gusher" or "spouter."

Noting the dangers of overexploitation, Assistant Secretary of the Interior Robert Herbst told Congress in 1979 that "unless geothermal exploration and development is very carefully planned . . . irreversible damage to . . . Yellowstone is a distinct possibility."

Yellowstone experts at the U. S. Geological Survey also advise caution, but consider "unlikely" any major movement of hot water between Island Park and the Yellowstone caldera. They propose an extensive monitoring system to allow geologists to predict any possible change in the geothermal system long before it could affect the Yellowstone geysers.

Concern about these fragile systems pales before the prospect of Yellowstone's next major eruption. The three most recent explosions "were on a scale not known in recorded history," says Dr. Robert L. Christiansen, a USGS expert on Yellowstone's volcanism. "If an eruption like that were to happen today. . . ." His voice trails off as his mind's eye gazes across a dozen western states buried in ash.

USGS geologists identify the Snake River Plain and Yellowstone as one of several lines of potential volcanic activity threading through the American West. They warn that any of these powder kegs could blow at any time.

Indeed, not just one but three major fault zones, each hundreds of miles long, intersect at Yellowstone, an indication of the great forces that must be stressing the North American Plate beneath.

The most dramatic proof of that occurred at 23 minutes before midnight on August 17, 1959, when the largest earthquake in the region's history hit near Hebgen Lake, six miles west of the Park. In less than a minute, 60 million tons of mountainside tumbled into the valley below, killing at least 28 people. The huge slide barely missed burying Rock Creek Campground.

In the park, most of the geysers and pools erupted simultaneously. Cascade Geyser, dormant for more than 50 years, literally blew its top; Giantess Geyser, normally active for 12 to 36 hours at a time, ran continuously for 4 days. More than 160 geysers with no record of eruption suddenly spurted to life.

Geologists have seen a less dramatic but equally disturbing sign that Yellowstone's violent past may repeat itself: Parts of the caldera floor are rising 14 millimeters (half an inch) every year, a sign that the magma chambers may be refilling.

If there is an eruption, we hope to have ample warning—an increase in hot-spring activity, dormant geysers that reawaken, and more frequent earthquakes, particularly the swarms of minor quakes that signal stresses in the earth's crust.

Geologists do not know what will happen beneath Yellowstone—or when. Always they come back to the same arithmetic: three eruptions, one every 600,000 to 800,000 years. And it's been 600,000 years since the last. "The history of any one part of the earth," said geologist D. V. Ager, "consists of long periods of boredom and short periods of terror."

underlying the Snake River Plain.
And yet the lava now covering
these old calderas is basaltic, and
flowed millions of years after this
area had passed over the hotspot.
Does Yellowstone mark the head
of a "propagating rift," a giant
rip in the continent that releases
basalt in its wake? That's one more
hotspot mystery to solve.

Slipping

OVERLEAF: *An eroded creek bed near California's Carrizo Plain follows the San Andreas Fault.*

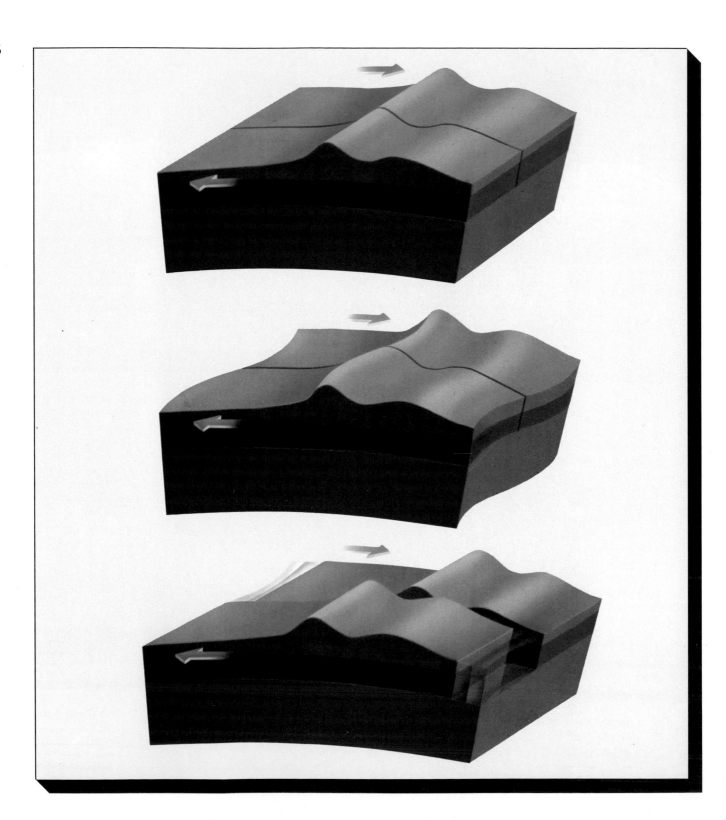

Slipping:
Earthquake Maker

An earthquake waiting *to happen—in the words of one geologist, it's "one big fire cracker indeed." Energy which has built up over years or centuries, to the force of one billion tons of TNT, blasts free in only a few moments.*

Diagrams show what occurs as two plates struggle to slide past one another along a fault, *a fracture in the crustal rock (opposite, top). If the rock along the fault locks, strain builds. Under stress from continued plate motion, the crust on each side of the fault slowly bends, or* deforms *(middle), storing up elastic energy like a spring wound tighter and tighter. Finally the spring lets go (bottom). Rock on one side of the fault jerks violently past the other side, sending out the shock waves of an earthquake.*

When plates move past each other horizontally—the motion is called translation—*crust is neither created, as at a spreading ocean ridge, nor consumed, as at an ocean trench. Geologists term the line of contact between the translating plates a* strike-slip fault. *Movement occurs sideways, along the* strike, *or horizontal axis, of the fault.*

The map highlights some of our planet's major translation zones. In some places, the plates move in opposite directions; in others, both plates move in the same general direction but at different speeds. Either way, they create strike-slip faults that threaten surrounding regions with earthquakes.

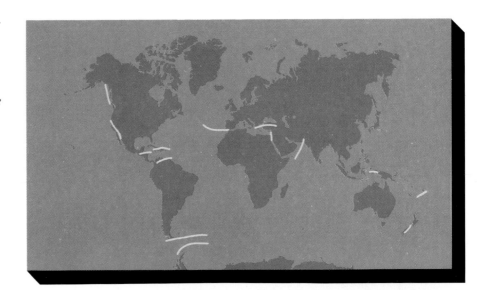

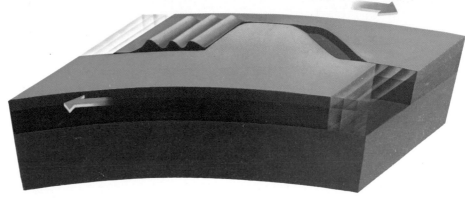

If a strike-slip fault bends, *a* pull-apart *basin may grow (above, right). The crust stretches thinner and thinner, like taffy in a taffy pull. Eventually it collapses, creating a long, narrow depression which may collect sediment or fill with water, as in Guatemala's Motagua Valley or in the Dead Sea basin between Israel and Jordan.*

If the fault bends the other way, plate motion jams the blocks of crust together, creating ridges and folds (above, left). California's Transverse Ranges, north of Los Angeles, have formed where a great bend in the San Andreas Fault wrinkles the crust. Just southeast of the mountain ranges lies the pull-apart we call the Salton Sea. Pull-apart basins and transverse ridges vary in size and age. They may be geologically complex, heavily faulted areas, with volcanism and hot springs.

On the Grindstone Edges

Half a minute's shudder of Earth leveled this Guatemalan mountain village. The town lay near the Motagua Fault, where Caribbean and North American Plates grind past one another. In 1976, after some 60 years of accumulating strain, the fault slipped in a lurch of destruction and death.

We rode on a sea of mountains and jungles, sinking in rubble and drowning in the foam of wood and rock. The earth was boiling under our feet . . . making bells ring, the towers, spires, temples, palaces, houses, and even the humblest huts fall; it would not forgive either one for being high or the other for being low. *Survivor of the 1773 earthquake in Guatemala*

The Motagua Fault is 40 million years old. It cuts across Guatemala, 240 kilometers (149 mi) long, 15 kilometers (9 mi) deep. As part of the boundary between the North American and the Caribbean Plates, the ground on the north side of the fault slides slowly westward past the ground on the south side. Sometimes it sticks.

At 3:02 a.m. on February 4, 1976, the tension broke. As people slept, the great rupture tore Guatemala in two, bringing the greatest natural disaster ever recorded in Middle America. Fragile adobe construction and quake tremors make a deadly combination. More than a million people lost their homes, 77,000 were hurt, 23,000 died. Adobe walls crumbled; wooden beams and tile roofs crashed to the ground in more than 300 villages.

For years before the earthquake, geologists had discussed the great fault system that scars the Guatemalan landscape. Was it a plate boundary or not? Geologist George Plafker said after the 1976 nightmare, "I hope this will convince some of the skeptics."

It was the extent and amount of surface movement that qualified the Motagua Fault as a plate boundary: A zone of cracks almost the full length of the fault ripped open, and in some places land on one side slipped 3.4 meters (11 ft) sideways past the other side—the greatest such jump since the 5.5-meter San Andreas slip in 1906. The behavior of the Motagua showed, stated Plafker, that it had broken under strain produced by forces which have been pushing the Caribbean Plate eastward for millions of years while squeezing its western corner between the North American Plate and the smaller Cocos Plate. In the process, Plafker suggests, the eastern part of Middle America may slide completely away from the western part—but that's millions of years ahead of us.

In the meantime, as Earth slowly remodels itself, plate boundaries that show translational movement create earthquake hazards in many places. One of the most studied regions is the San Andreas Fault zone in California. Across New Zealand, the Alpine Fault slices a gentle curve. In Turkey, the Anatolian Fault system fractures the crust where the small Anatolian Plate squeezes past the Eurasian Plate on the north. Of the Levant Fault zone that runs under the long-chronicled Dead Sea area, geophysicist Amos Nur says, "We have a historical record here of at least 3,000 years of earthquake activity."

"Joshua fit the battle of Jericho"

The oldest city in the world, Jericho was a great fortress in the fertile Jordan Valley. Joshua's army had to overcome the city walls after the Children of Israel crossed the Jordan River to the Promised Land. The Book of Joshua (6:20) tells the story: *And it came to pass, when the people heard the sound of the trumpet, and the people shouted with a great shout, that the wall fell down flat, so that the people went up into the city, every man straight before him, and they took the city.*

Scholars and archaeologists have long been fascinated by the story of Joshua's trumpet. In the early 1900s a German-Austrian expedition dug for clues. They found concentric fortifications ringing the city. An inner wall was almost 4 meters (13 ft) thick and 10 to 12 meters high, an outer one almost 2 meters thick and 7.5 to 9 meters high. Clearly it took more than a shout to topple these massive structures.

John Garstang, an archaeologist with a British expedition of 1930, found charred wood, ashes, broken stones, and other evidence that the city had burned in a great fire, just as the Book of Joshua says. Garstang also discovered earthquake evidence in Jericho's fallen walls. When stone walls collapse in tidy rows, the pattern indicates that an earthquake was responsible.

According to Amos Nur, earthquakes in this area have often caused mudslides that dam up the Jordan River for a day or two. Researchers have documented such events at least three times in the last 150 years; the last major quake here, which damaged Jerusalem and Bethlehem in 1927, dammed the river's flow. Thus, says Nur, the miracle that halted the Jordan so Joshua and his troops could cross was probably an earthquake.

When I looked over the site of Jericho I could see why. The Jordan Valley and the Dead Sea lie inside the Dead Sea Rift, an extension of Africa's rift system. As the separating African and Arabian Plates opened this rift millions of years ago, translational motion began to slip one side of the valley sideways past the other. Later, world sea levels rose, and water flooded in to form a long lake, Lisan. When the sea receded, the lake shallows dried up, leaving the Sea of Galilee, the Dead Sea, and the Gulf of Aqaba.

The Book of Zechariah (14:4) seems to describe fault movement: *And the mount of Olives shall cleave in the midst thereof toward the east and toward the west, and there shall be a very great valley; and half of the mountain shall remove toward the north, and half of it toward the south.*

This is what plate tectonics says. The Arabian Plate on the east side of the fault is moving north with respect to the African Plate on the west. This motion began about 15 million years ago. Rock formations once connected now stand as much as 110 kilometers (68 mi) apart.

Fire and brimstone

Then the Lord rained upon Sodom and upon Gomorrah brimstone and fire . . . out of heaven. . . .

But [Lot's] wife looked back from behind him, and she became a pillar of salt. . . .

And [Abraham] looked toward Sodom and Gomorrah, and toward all the land of the plain, and beheld, and, lo, the smoke of the country went up as the smoke of a furnace (Genesis 19:24-28).

This was perhaps 2000 B.C. The cities of Sodom and Gomorrah, notorious in history, may lie buried at the bottom of the Dead Sea. People have looked, but no one has yet

Lowest body of water on Earth, the Dead Sea lies in a pull-apart basin opened by a slight bend in a fault zone. The land has subsided here, drawn thin by the motion of the Arabian Plate as it slides past the slower-moving African Plate.

This plate boundary follows the Dead Sea Rift, which formed when Arabia split away from Africa about 15 million years ago. Little spreading occurs here now, but translational movement shakes the Bible Lands with frequent earthquakes and leaves many intriguing questions: Do the ruins of Sodom and Gomorrah lie beneath the salt at the bottom of the sea? Did earth tremors part the waters of the Jordan for Joshua?

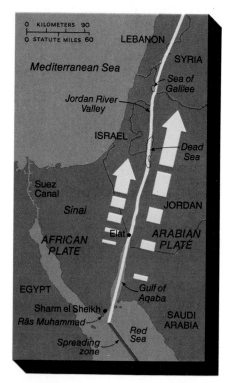

found them. Fire and brimstone? A pillar of salt? What could account for such stories?

Visiting the Dead Sea, I wondered why anyone would want to live on its shores. In summer, the sun bakes the region to over 40°C (104°F). This landlocked lake, 399 meters (1,309 ft) below sea level, evaporates quickly in the intense heat, and is about ten times saltier than the oceans—so salty, in fact, that people cannot sink.

In A.D. 70, the story goes, the Roman commander Titus ordered that a group of slaves be chained together and cast into the Dead Sea. But the men bobbed like corks. Each time they were thrown into the water, they drifted back to shore, until the amazed Titus spared their lives.

Pits of slime

Near the west shore of the Dead Sea, I smelled sulfur from the abundant hot springs. Genesis 14:10 says that when the kings of Sodom and Gomorrah tried to flee from here during a war, "the vale of Siddim"—the Dead Sea—"was full of slimepits," or natural asphalt, a bituminous substance. In fact, the Romans called the Dead Sea "Lacus Asphaltites."

After the earthquake of 1834, a raft of asphalt drifted ashore in the south end of the sea, so big that local Arabs took three tons of the stuff to market, where it was valued for use in medicines and in making cement and mortar. Three years later, after another earthquake, a mass of asphalt the size of a house rose to the water's surface.

After an earthquake such seepages of asphalt, petroleum, and natural gas near the presumed sites of

Sodom and Gomorrah could have caught fire, according to U. S. geologist Frederick G. Clapp, ignited by cooking fires or by lightning, and "would adequately account for recorded phenomena"—which is to say, destruction of the cities by fire and brimstone.

In flight from Sodom, Lot and his family heard the warning of God's angels: "Look not behind thee" (Genesis 19:17). But Lot's wife cast a farewell glance on the blazing city and, for her disobedience, was changed to a pillar of salt.

Most scholars believe that this legend is linked to a large mound of salt called Mount Sedom, a low plateau some eight kilometers long that flanks the sea on the southwest. Mount Sedom is a salt diapir, a protrusion which has pushed upward along the fault.

In 1848, leading a U. S. expedition down the Dead Sea, Lt. W. F. Lynch reported, "To our astonishment, we saw . . . a lofty, round pillar, standing apparently detached from the general mass. . . . We found the pillar to be of solid salt, capped with carbonate of lime."

What Lynch saw was a part of the mountain, shaped to human form by erosion. Runoff and erosion repeatedly create and destroy "Lot's wives" from the ribs of Mount Sedom.

The Dead Sea lives

For the last 50 years or so, the level of the Dead Sea has been falling. As I drove along the western shore between Jericho and Masada, I could read a drop of at least two meters recorded by recent water lines along the banks.

Every year the sun evaporates

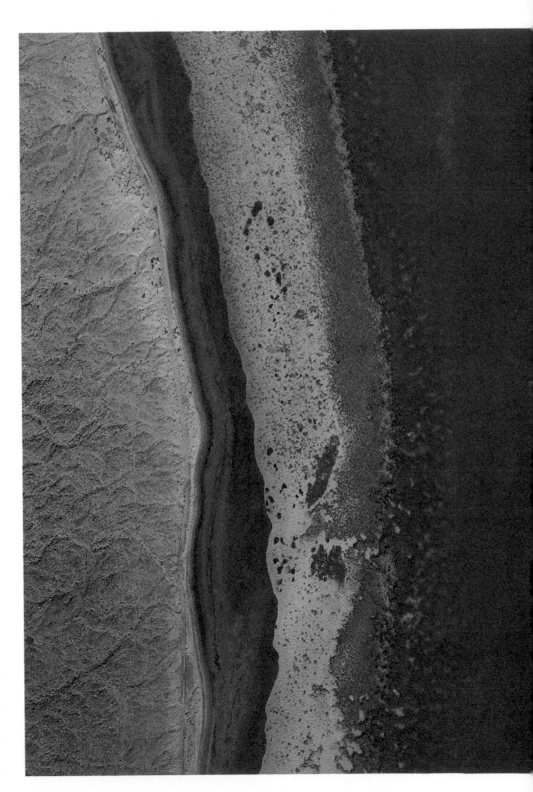

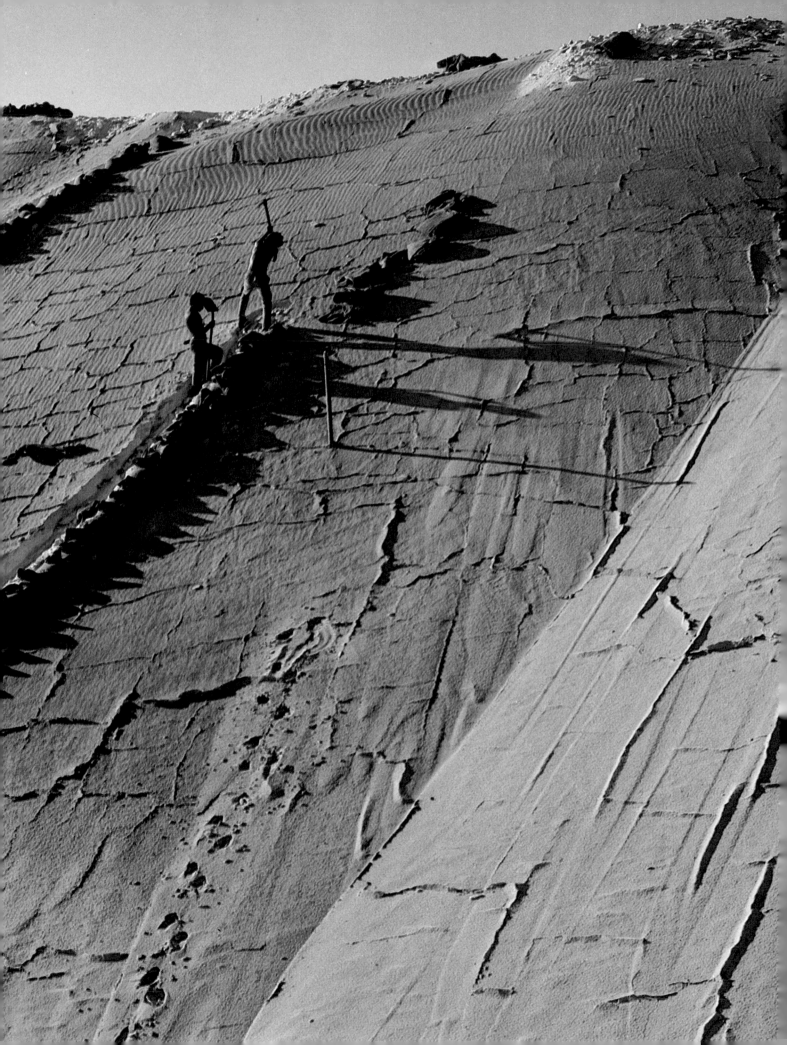

A small mountain of potash rises on the southwest shore of the Dead Sea. Laborers at the Dead Sea Works, near Sedom, dig trenches and fill sandbags to anchor protective plastic covers. Potash, a key ingredient in fertilizer, precipitates from this unusual lake—a mineral-rich brine ten times saltier than ocean water.

Bromine, glowing amber as a gas, then red as a condensed liquid, undergoes processing at the Dead Sea Works. Used in leaded gasoline, medicines, dyes, and photographic film, bromine exists in the bitter Dead Sea water in combination with other minerals. Scientists believe the Dead Sea salts have ancient marine origins.

more than one billion cubic meters of water from the Dead Sea, leaving a residue of valuable minerals. In times past, the water lost to evaporation was replaced by underground springs and incoming rivers. Most of the new water originated at one source, the Jordan River, at the north end of the sea.

But today the Israelis have diverted the Jordan in order to supply water for a growing population. In addition, the Jordanian government has begun to divert tributaries. In a few years the Jordan may be a dry creek bed.

Already the Dead Sea is low enough to isolate the southern basin from the deeper northern basin and thus jeopardize an important industrial operation. The Dead Sea Works, at the shallow southern end, annually extracts by solar evaporation over one million tons of potash from the salty water. Because of the lowered water level in the south, the Israeli government has built a canal connecting the evaporation ponds there to the mineral-rich water in the north.

The Israelis have also responded to the prospect of a continued drop in water level with a plan that gets more complicated each year. I visited the headquarters of the works at Newe Zohar and talked with Information Officer Shlomo Drori.

At the heart of the plan, he said, is a power plant and an 800-million-dollar, 100-kilometer-long (62-mi) channel between the Mediterranean Sea and the Dead Sea—about 80 percent of it tunneled through mountains. A real challenge has been to find as many uses as possible for the expensive waterway.

One use depends on the success

of experimental solar energy ponds. For five years or so, the ponds can use Jordan River water. Then the Mediterranean-Dead Sea Water Channel can take over as the Jordan's flow declines. Channel proponents argue that it can also help produce safe nuclear energy. It will allow atomic power plants, which need a steady water supply for cooling, to be sited inland next to Channel-fed artificial lakes, instead of in their present locations along the densely populated, militarily vulnerable Mediterranean coast.

On the east bank of the Dead Sea, Jordanians too are making use of the brine with a 425-million-dollar project, the Arab Potash Company. Officials at the Dead Sea Works speak hopefully of cooperation between the two enterprises, so that both can enjoy Earth's bounty.

"The damndest finest ruins"

The animals were the first to sense the danger. Dogs began to bark, whine, and run about. One jumped out a window. Snorting firehorses broke out of their stalls at the San Francisco Fire Department.

Soon people heard a low rumbling "like distant thunder." They heard "a noise in the trees as though heavy wind were blowing through them." A farmer said he felt a wind in his face.

The earth along the San Andreas Fault was moving. In the town of Saratoga, a woman reported waves rippling across the ground, "the orchard trees rising and falling on each wave, like ships at sea, while the electric poles along the road leaned this way and that, some

Salt and Stone: Bible Land Geology

Small pillars of salt glitter near the shore of the Dead Sea, attesting to the Hebrew name for this briny pool, Yam Hamelach—"sea of salt." These crystal structures grow above the waterline to a maximum height of about half a meter, built up by a wicklike property of the salt. The growth starts when a stick or a piece of wood juts from the water. Then evaporation by a blazing sun dries up the water. 219

None of these little columns will get big enough to stand in for the biblical pillar of salt that Lot's wife supposedly became when she disobeyed God's commandment. Those pillars are protrusions formed now and then by erosion on nearby Mount Sedom, a plateau consisting largely of salt.

OVERLEAF: *Masada, an ancient, ravaged rock prow in the Dead Sea basin, was a desert fortress 2,000 years ago. Archaeologists have excavated large portions of the stronghold where almost 1,000 Jewish Zealots withstood a two-year Roman siege, then chose mass suicide to avoid capture. Partially restored living quarters stand on the upper terrace; in the foreground Roman ramparts confront the butte.*

Archaeologists discovered telltale recurring patterns in the rubble of numerous collapsed stone walls, graphic evidence of the earthquakes that have shaken Masada since its days of glory.

310 million years ago

Fault

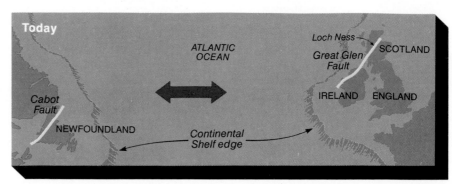

Today

Loch Ness
Great Glen Fault
SCOTLAND

ATLANTIC OCEAN

Cabot Fault
NEWFOUNDLAND

IRELAND
ENGLAND

Continental Shelf edge

Loch Ness, on Scotland's Great Glen Fault, has a past stranger than any fabled beast that may haunt its inky depths. Geologists now believe that the other end of the Great Glen lies 5,000 km (3,100 mi) away—Newfoundland's Cabot Fault. Some 400 million years ago, when Europe and North America were part of one continent, a strike-slip fault sheared across the future islands of Newfoundland, Ireland, and Great Britain (far left). Plate movement later rotated them slightly. When the Atlantic Ocean opened some 200 million years ago (along the red line), the old fault broke in two. Spreading still widens the gap today (left).

OVERLEAF: *Tourists frolic in water heated deep in the earth at Pamukkale, Turkey. The hot springs, south of the Anatolian Fault, bring dissolved minerals from deep in the lithosphere, then deposit them to build up terraces. An earthquake in 1354 first set the waters free from their hilltop course to cascade over the ledges.*

223

seeming almost to touch the ground." Some witnesses thought that the waves rose a meter in height and spanned 20 meters from crest to crest.

The quake had struck. It was 5:12 on Wednesday morning, April 18, 1906. For 45 seconds California experienced some of the worst earth tremors in U. S. history. People, animals, and buildings were thrown to the ground. In some places, loosely filled "made land" turned to mush, tipping buildings intact.

A man from San Jose reported, "About the middle of the quake, these [waves] were met by other waves and the whole surface resembled hillocks or cross seas, and the tree-tops waved wildly." The sensation of swaying ground made some people sick.

Then the fires began to break out, ignited by kerosene lamps and wood-burning stoves. People said later that these fires, not the actual quakes, took most of the 700 or so lives lost in San Francisco and nearby cities. The fires did most of the damage to buildings.

Just six months earlier, the National Board of Fire Underwriters had marveled that San Francisco had not already burned to the ground. Now, for three days, the flames licked through its streets. The fire department tried in vain to contain the blaze at the waterfront, site of the first big outbreaks. Here at least was a source of water for a city where most of the mains had broken in the earthquake.

Men who had never used dynamite before set off charges, trying to build a firebreak. The attempt did more harm than good. In just one hour the dynamite squad started five new fires. Explosions turned some buildings into great piles of kindling wood. They sent aloft blazing timbers which fell onto houses that had, so far, escaped the flames. People across the bay in Oakland watched a great cloud of smoke darken the sky. "The sun," said earthquake refugee Mary Austin, "showed bloodshot through it as the eye of Disaster."

In an Italian neighborhood people met the flying sparks by carrying several hundred thousand gallons of wine to their rooftops, where they poured it out to dampen the shingles. Their houses and stores survived.

By Saturday, the 21st, San Francisco had sustained half a billion dollars in damage. Almost 3,000 acres lay—as poet Lawrence W. Harris boasted with civic spirit—in "the damndest finest ruins."

Santa Rosa, San Jose, Agnews, and other towns near the San Andreas saw proportionately much greater earthquake damage. At the site of greatest horizontal slippage—as much as 5.5 meters (18 ft)—modern Californians trudge an Earthquake Trail in Point Reyes National Seashore and ponder photographs of the 1906 destruction.

In Palo Alto, on the peninsula south of San Francisco, the sandstone buildings of the 15-year-old Stanford University stood on the gravelly soil that forms the Santa Clara Valley floor—a site shakier than bedrock but more stable than the mud and saturated soils closer to the bay. A dormitory, the library, and the gymnasium were wrecked. The stone tower of the church crashed through the roof. In the church, shattered art treasures

littered the floors. The great stone gates collapsed in a heap of rubble. The quake toppled the statue of Louis Agassiz, famed naturalist and geologist, and left it sticking head-first in the pavement below.

The new three-story building that housed the Geology Department was empty when the tremors knocked down parts of every wall. Years later I would teach in the reconstructed building. My office window was under a cornerstone on which was carved GEOLOGY. Agassiz's fate led me at times to wonder, if another quake should hit, whether I might have GEOLOGY for my tombstone.

The next one?

Geologists say that a snag on the northern San Andreas lasts about 125 years, on the average, between big jumps. Dr. Kerry E. Sieh, an expert at the California Institute of Technology, maintains, "There's a 50 percent chance of a great quake in the next four decades." And that could mean tomorrow! The real question is not *if* but *when, where,* and *how big*.

Despite predictions like Sieh's, life in California goes on. People and industries continue to move into the state and build their lives right along the very faults that scientists say could release unimaginable amounts of destructive energy. Why do they do this?

The answers are as complex as the faults. Partly it's because an earthquake threat, to the average person, is silent and invisible. And it's partly because earthquake prediction is a new science. We can't give forecasts the way the weather people can.

Quakes east and west

For the first 25 years of my life I lived in southern California. My family grew accustomed to the occasional tremor, the swaying of a chandelier. We would joke about keeping the best bottle of bourbon safely at the back of the shelf. And we knew something of California's history: that in 1857 the greatest slip ever on the San Andreas—some ten meters—took place on the sparsely populated Carrizo Plain northwest of Fort Tejon; that in 1872 the Owens Valley Fault zone shook at least as hard as San Francisco had in 1906, maybe harder; that since 1800 at least 35 earthquakes of 5.3 and up on the Richter scale had shaken the state.

When my wife and I moved to the East Coast in 1967, we didn't *feel* any safer from quakes. On the average, instruments record hundreds of earthquakes a day all over the United States. Only one or two are strong enough for people to sense them. Over the years, Alaska, California, and Hawaii have had the most tremors. But in states as far apart as Wyoming and Virginia, people report feeling shakes.

From the earliest settlements along the eastern seaboard, people have recorded local tremors. A Plymouth, Massachusetts, Pilgrim, in June 1638, wrote of "a great and fearfull earthquake" which "caused platters, dishes, & such things as stoode upon shelves, to clatter and fall downe."

The years 1811 and 1812 saw some of the most severe shakings in U. S. history center on New Madrid, Missouri. (Davy Crockett was there soon after and named it the "Shakes Country.") Three big

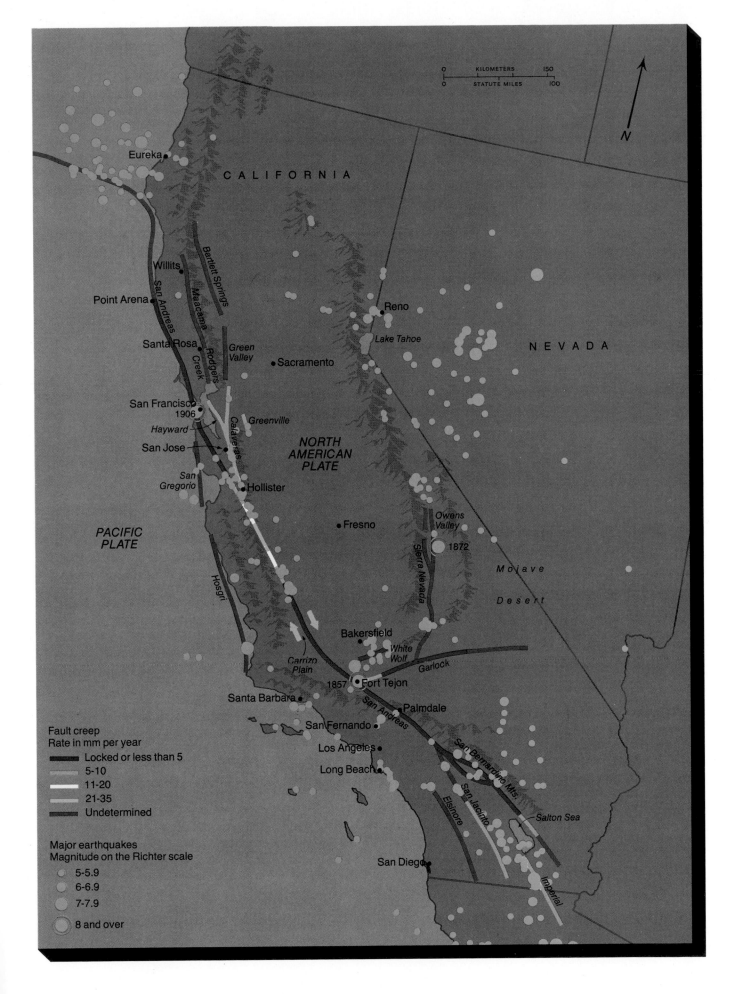

227

Fault creep
Rate in mm per year

Locked or less than 5
5-10
11-20
21-35
Undetermined

Major earthquakes
Magnitude on the Richter scale
5-5.9
6-6.9
7-7.9
8 and over

CALIFORNIA

NEVADA

PACIFIC
PLATE

NORTH
AMERICAN
PLATE

Eureka

Willits

Point Arena

Santa Rosa

San Francisco
1906

Hayward

San Jose

San
Gregorio

Hollister

Bartlett Springs

Maacama

Rodgers
Creek

Green
Valley

Greenville

Calaveras

Reno

Lake Tahoe

Sacramento

Fresno

Owens
Valley

1872

Sierra Nevada

Mojave

Desert

Hosgri

Carrizo
Plain

Santa Barbara

San Fernando

Los Angeles

Long Beach

Bakersfield

White
Wolf

Garlock

1857 Fort Tejon

San Andreas

Palmdale

San Bernardino Mts.

Elsinore

San Jacinto

Salton Sea

Imperial

San Diego

The leaps and bounds of an earthquake begin at the focus, where a locked fault suddenly lets go. Shock waves spread outward. Primary, or P, waves, speed along at 5.5 to 8.5 km per second (12,000 to 19,000 mph), to arrive first at the surface and at seismometers. P waves are compression waves, alternately pushing and pulling surface structures in the direction of wave travel. Secondary, or S, waves are slower, 3 to 4.6 km per second. They arrive next at the seismometer, heaving the ground up and down or from side to side. S waves cannot move through water. A third type, surface waves, produce low frequency vibrations that roll the ground or whip it sideways. Since these travel more slowly and take longer to diminish, they can sway tall buildings at great distances from the epicenter.

Because P waves travel like sound waves, vibrating back and forth, they set the air in motion when they leave Earth's surface and prompt reports that the quake "roared like a locomotive."

228

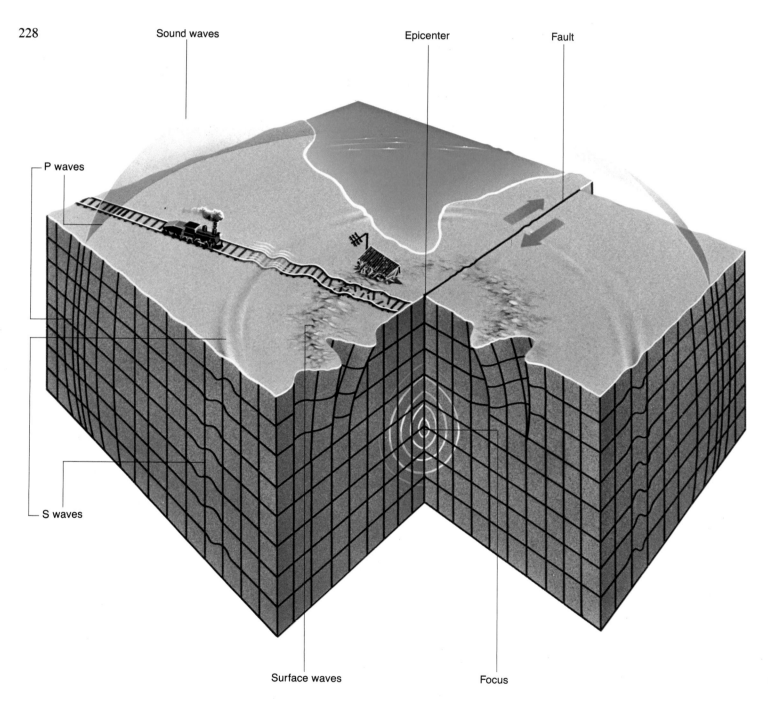

Sound waves

Epicenter

Fault

P waves

S waves

Surface waves

Focus

quakes struck near a meander loop of the Mississippi River where Kentucky, Missouri, and Tennessee come together. Their magnitudes have been estimated at 8.6, 8.4, and 8.7. People felt ground shaking in an area that stretched from Oklahoma to Virginia, and from southern Canada to southern Louisiana.

This was Chickasaw land, with few settlers. But the first quake convinced some they'd awakened to Judgment Day. Geysers of sand and water blew to the sky. Earth waves swelled, islands disappeared into the river, the air smelled like sulfur. With the third quake, in February 1812, one witness said, "The Mississippi seemed to recede from its banks . . . its waters gathering up like a mountain." When it was all over, the course of the river had been changed, and there was a new lake in Tennessee, 30 kilometers (19 mi) long—Reelfoot—where today's vacationers play.

Charleston, South Carolina, suffered an earthquake in 1886 that killed 60 people. In the 20th century, northeastern North America has shaken more often than most people realize, with tiny quakes taking place regularly in many areas, and bigger ones tumbling chimneys from New York (Attica, 1929) to New Hampshire (Ossipee, 1940), and registering as high as 5.9 in January 1982 for a tremor centered in New Brunswick, Canada.

These earthquakes do not occur along plate boundaries. But U. S. Geological Survey researchers believe that they may have found fault-zone behavior that would explain them. In Tennessee, for example, there are buried faults that may have been squeezed whenever the North American Plate ground hard against another plate and stretched during periods when continents were splitting apart. But the picture is complicated and the source of present-day stress is uncertain. The faults under New Madrid seem to be nearly vertical (see page 63), while beneath Charleston a fault underlying a thin, horizontal sheet of rock may be to blame for that city's destructive quakes.

So long, Los Angeles
On the average, over the 1,600-kilometer (1,000-mi) length of the San Andreas Fault system, a spot of ground on one side of the fault shows two or three centimeters a year displacement, relative to a spot opposite. It will be a long time before the Dodger-Giant baseball game is a crosstown rivalry, but in a million years translational movement has brought the sites of Los Angeles and San Francisco 50 kilometers closer together.

Farther back from the fault, San Diego sits on the western block, the Pacific Plate, and Oakland on the eastern, the North American Plate. If the present rate of movement continues, the two will pass about 15 million years from now. Since the fault extends into the Gulf of California, in 20 million years more Baja California will be cruising past San Francisco Bay.

In the southern Coast Ranges, 175 kilometers (109 mi) south of San Francisco, Pinnacles National Monument raises a spectacular skyline of light red, green, and gray lava 23.5 million years old, a remnant of Miocene volcanism.

The pinnacles lie west of the San Andreas on a steadily creeping

"The trouble began in the sea"

It ripped out of the water, following the San Andreas Fault, added California naturalist David Starr Jordan, like a "devouring dragon, leaving its trail on the hills and destroying the works of man." It smashed some buildings to smithereens, tipped others crazily, like the houses on Golden Gate Avenue in San Francisco (left). For 45 seconds on that April daybreak in 1906, the earth shook. Even before the tremors died, the fires began to blaze. By next morning, flames had swept away this block.

OVERLEAF: *The tumbled wooden cupola of the Santa Rosa Courthouse lies on a roof of the three-story brick building. Although 32 km (20 mi) separate Santa Rosa from the fault, the town suffered greater earthquake damage for its size than any other. Almost the entire business section collapsed because of poor mortar and faulty construction.*

2ND OVERLEAF: *From Telegraph Hill, survivors contemplate a San Francisco devastated by four days of fire and a moment's slip of the earth along a crack 20 million years old.*

231

stretch of the fault. Volcanic rocks east of the fault match those in the monument, but, now buried under smooth hills, they lie where they formed, over 300 kilometers (186 mi) south. The western piece of crust has drifted that far since the Miocene volcanism.

Geophysicist Tanya Atwater has suggested that the effects of this motion have not all been absorbed by the San Andreas system. They may reach as far as the desert regions of Mexico and the western U. S. known as the Basin and Range. If they do, the boundary between the two plates is not a clearly defined fault, but a broad, fragmented zone which may be linked to an increasing number of earthquakes and volcanic rumblings on the California-Nevada border.

A separate peace

A good place to look at slow creep in action is the town of Hollister, which lies along a fault related to the San Andreas, the Calaveras Fault. Sensitive seismographic instruments in the Hollister area pick up 20,000 earthquakes a year, mostly very tiny ones. In 1961 there was a jolt of 5.6—the biggest since a 6.6 in 1911. Yet, as I drove across the valley in a comfortable car, enjoying music from the radio, I could understand how difficult it is for most people to realize the danger that lurks under the ground.

In Hollister I saw people pushing baby carriages down San Benito Street, seemingly unconcerned about the Calaveras Fault grinding its way through town.

George Curtis, a Hollister resident since 1935, was nonchalant. He owns two adjacent houses on

Fifth Street, a peaceful residential area in the middle of town. The fault runs between the houses. In front of them, a white picket fence makes a wide offset bend near its midpoint.

"I put that fence in about 15 years ago and it was straight then. Now it's bent about three to four inches. Every weekend, tourists come by, look at the fence, and ask me, 'What's it like to have a fault run under your house?' I tell them it doesn't bother me a bit. I've lived here most of my life and I haven't gotten a bloody nose yet."

George's houses stand about 12 feet apart. One is an old Victorian-style building on a redwood foundation which "just gives." The other has a badly cracked concrete foundation. The curb in front is broken and offset a couple of inches where it intersects the fault. Continuing across Fifth Street, the fault offsets the white centerline, which periodically has to be repainted to make its two ends meet.

Across the street, an elderly woman stood sweeping along an offset curb. The fault trace passes through her yard, beneath her house, and enters a small park where children play on the slope of a low scarp.

The fault is a part of normal life. Its creep and small-to-moderate quakes can be annoying, occasionally do some damage, but do not seem to be a serious threat to life. In Vieire's Liquor Store on the corner of San Benito and Hawkins Streets, wires stretch past the shelved bottles to keep them from falling to the floor during an earthquake. I asked Rita Fernandes, the owner, if her insurance requires this protection.

"No, we don't have earthquake

Home on the Ranging Plate

The seaward slice of the San Francisco Bay area is a newcomer to the neighborhood. This piece of resistant granitic crust began its journey from the south about 20 million years ago as part of the Pacific Plate. It has ground its way northwestward, pushing and slipping against the rock of the North American Plate.

The seam is the San Andreas Fault zone, the longest of the many faults in California. Here in the bay area, the San Andreas is locked. At more freely moving points along the fault, creep averages 3.5 cm per year. On the Hayward Fault—part of the larger system—creep is much slower.

But a major earthquake can bring a leap of 5.5 m (18 ft), perhaps more, as San Franciscans saw in 1906. Some four million people live between the two faults, many in houses built on sagging, slipping ground crushed by fault movement or on unstable landfill.

The Hayward Fault could let go at any time, producing an earthquake of magnitude 7.0 to 7.5. If the San Andreas repeats its 1906 performance (8.3)—as it could in the next 40 or 60 years—bay area residents might see losses of 20 to 30 billion dollars from shaking alone, in addition to fires, landslides, and other catastrophes. Thousands of deaths and hundreds of thousands of injuries could result. Both citizens and scientists try to plan for the quake that is sure to come.

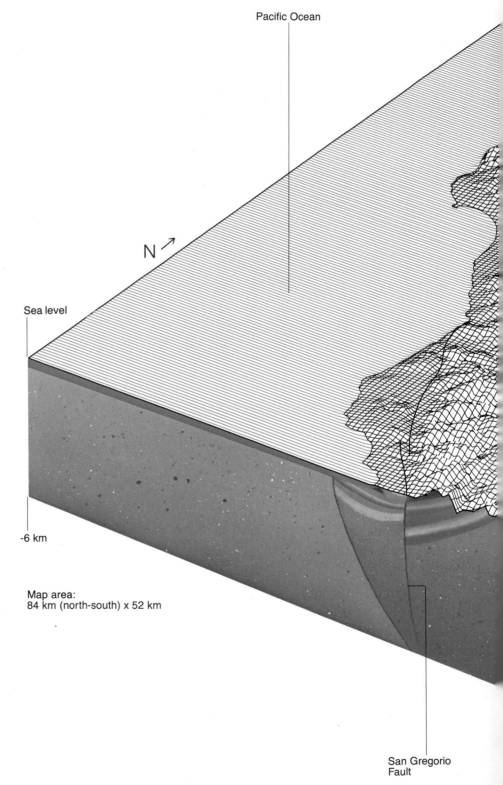

Pacific Ocean

N ↗

Sea level

-6 km

Map area:
84 km (north-south) x 52 km

San Gregorio
Fault

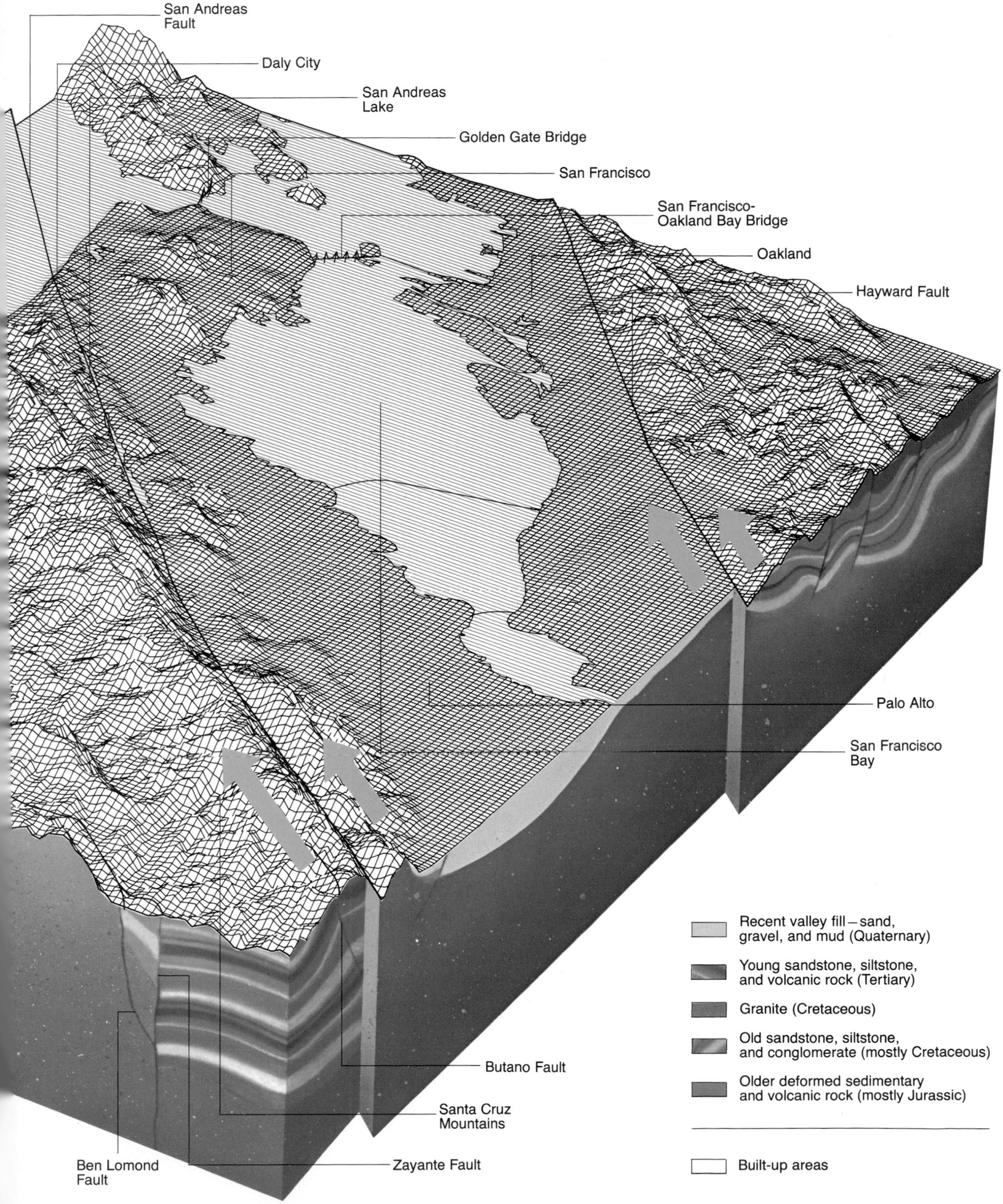

San Andreas Fault

Daly City

San Andreas Lake

Golden Gate Bridge

San Francisco

San Francisco-Oakland Bay Bridge

Oakland

Hayward Fault

Palo Alto

San Francisco Bay

Butano Fault

Santa Cruz Mountains

Ben Lomond Fault

Zayante Fault

Recent valley fill—sand, gravel, and mud (Quaternary)

Young sandstone, siltstone, and volcanic rock (Tertiary)

Granite (Cretaceous)

Old sandstone, siltstone, and conglomerate (mostly Cretaceous)

Older deformed sedimentary and volcanic rock (mostly Jurassic)

Built-up areas

Accommodations to life in California earthquake country: Bob McJones levels his house in Palos Verdes with jacks, as the unstable soil beneath it slowly slides into the ocean; restraining wires serve as do-it-yourself earthquake insurance for Vieire's Liquor Store in Hollister.

insurance," she replied. "If we lose any bottles, we just lose them." She explained that when the last earthquake took place, "the bottles just shook from side to side. They don't jump out like in some other liquor stores down the street, where the shelves run the other way."

I saw a dramatic example of slow creep at Almadén's Cienega Winery, south of Hollister. In the 1940s scientists put notches in the freshly poured concrete floor of a storage area. The gradual offset has shown that one half of the room is moving slowly past the other half. The continuous creep has offset a drainage ditch on the south side of the building at least a foot, and cracked the north wall and bent it just like George Curtis's fence. On the winery grounds, I saw a man busy at yard work. I asked if doing his job on an active fault caused him any concern.

"Nope!" he replied. "What can you do about it?"

But such a truce with the faults cannot be called everywhere in California. Where a fault locks, the pressure begins to build. Eventually it reaches the breaking point. The plates snap past one another like a released spring. It's been called "instant creep," since movement that might have required decades of creep takes place in seconds. In other words, a major earthquake.

How big was it?

Newspapers usually report an earthquake's magnitude on a scale developed in the 1930s and '40s by Charles F. Richter and Beno Gutenberg of the California Institute of Technology. We who have lived most of our lives in California have seen many small and moderate quakes that measure around 5 or 6 on the Richter scale. So we may not worry much about Richter 7, 8, or 9. We should.

The Richter scale is logarithmic. Each increase of 1 stands for an increase of 32 times in the energy of the tremor and of 10 times in the size of the seismic wave measured by a seismograph. This means that an earthquake of magnitude 8.3—such as the one that hit San Francisco in 1906—is not twice as big as one of, say, 4.2. It is one million times bigger. Although there were no seismographs at Fort Tejon in 1857, modern estimates match the magnitude of that quake to the great one in San Francisco. The 1961 Hollister earthquake of 5.6 was "moderate." All it did was deliver minor damage to more than half the buildings in town.

In 1902 Giuseppe Mercalli, an Italian seismologist, devised a system that ranks tremors according to eyewitness accounts. Quakes range from those that are hardly perceptible, assigned the number I on the Mercalli scale, to number XII shakings, so severe that rivers can alter their courses, large rocks are wrenched loose, objects are thrown into the air, and almost all buildings and other structures collapse.

Surviving the big one

Many earthquake watchers list the southern San Andreas, a region that includes Los Angeles, as one of the most probable sites for a great quake in the next 30 years.

The San Francisco earthquake of 1906 told us a lot. For example, it showed that frame buildings stand up much better than brick or block

Atomic clocks, the speed of light, and space technology join the attempt to forecast earthquakes. Even small crustal movements show up on a range of instruments. NASA radio telescopes pick up signals from quasars at the edge of the known universe. As the signals arrive at each of two ARIES antennas, atomic clocks measure the tiny differences in arrival time. Data show changes in latitude, longitude, and elevation where the antennas stand on either side of a fault. Scientists study and correlate readings, on the lookout for earth behavior that may warn of an earthquake.

Geologists (below) operate an ultraprecise laser-ranging device.

244 or stone. Wood frames sway with the shock. Masonry can't unless it is well reinforced; it may collapse. In the 1906 tremor, buildings constructed on rock or solid ground did well, but those on landfill did not.

They fell prey to a phenomenon called liquefaction. When landfill or soft ground is violently shaken, the groundwater confined in porous layers turns the soil briefly into quicksand. The ground flows and destroys foundations. Buildings sink or tip over. As shaking subsides, the ground sets again, locking the buildings askew.

Although many of the built-up areas of California rest on unstable ground, the state has done some things that will lessen damage in the next big earthquake. New laws have strengthened building codes. The Field Act, which followed the 1933 Long Beach quake, set up strong standards for school construction. The possibility that the Van Norman Dam could have collapsed during the 1971 San Fernando earthquake led to closer dam inspection. The Alquist-Priolo Act of 1972 forbids construction intended for human occupancy within 50 feet of the trace of an active fault.

Even so, some modern buildings have little chance of surviving a major quake. "Tilt-ups" are made of prefabricated concrete slabs taken to a site—an industrial park, for instance—and tilted upright to be fastened to a roof. Earthquake shaking wrenches the connections loose and the building falls apart. Another problem is slab-floor construction—buildings designed like stacks of concrete pancakes, one slab on top of another. In a 1967 quake in Caracas, Venezuela, a building like this collapsed, killing 47 occupants in a deadly accordion.

Even if all future construction in California follows new building codes, many old, unreinforced masonry buildings still stand, some 8,000 of them in Los Angeles alone. Most are in run-down sections of town. If a great earthquake were to strike in Los Angeles tomorrow, 80 percent of the deaths possibly would occur there, where the poor live and work.

Leapin' lizards!

The pigs are climbing the walls, the chickens won't go in the chicken coop. Trained police dogs run around howling and sniffing the ground. Hibernating snakes crawl from their warm hollows to freeze on snowy fields. Frogs jump out of lakes through holes in the ice.

All this was happening in Liaoning Province in northeastern China during the winter of 1974-75.

Scientists watched this strange animal behavior. Seismologists, monitoring their instruments, saw a series of small earth tremors and a drop in the groundwater level. On the morning of February 4, tremors increased. Around 2 p.m. officials began broadcasting evacuation orders. Perhaps three million people left their homes and workplaces for makeshift shelters in the bitterly cold countryside. They waited until 7:36 that evening, when an earthquake measuring 7.3 racked the province. Devastation was widespread, and Haicheng turned to rubble. But of the enormous population in the region, only a few hundred died. Estimates are that tens of thousands of lives were saved, partly by listening to the animals of Liaoning Province.

Rodent radar

Cockroaches, chimpanzees, catfish —scientists all over the world have been studying them to answer the question: Can animals sense an impending quake? If so, we may be able to design instruments that can sense the same earthquake precursors that the animals react to.

Abnormal conditions in the environment may be the clue. Animals might sense an increase in the radon gas released into groundwater from rock under stress. They might hear high frequency noise produced by small foreshocks. Restless or disoriented animals might be detecting ions, electrically charged particles, released from strained rock, or they might be reacting to related changes in the static electrical field naturally present in the atmosphere.

Chinese experimenters have discovered that pigeons have delicate nerves in their legs that pick up minute vibrations in the ground. Because of reports that deepwater fish move into shallow water just before an earthquake, Japanese scientists have monitored catfish in special tanks. In California, dog breeders, housewives, ranchers, and zoo keepers formed a network they call "Project Earthquake Watch," which reached from Los Angeles north to where the San Andreas goes out to sea. They made detailed reports of their animals' behavior. But they learned little about forecasting because none of the accounts coincided with what project director William Kautz called "the right earthquake."

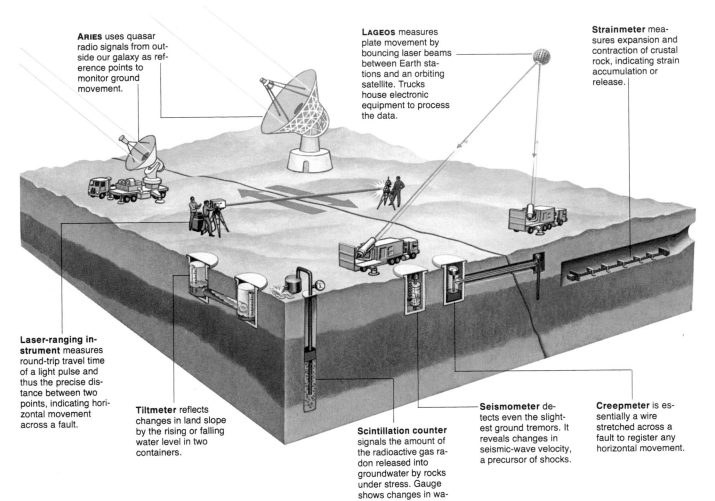

ARIES uses quasar radio signals from outside our galaxy as reference points to monitor ground movement.

LAGEOS measures plate movement by bouncing laser beams between Earth stations and an orbiting satellite. Trucks house electronic equipment to process the data.

Strainmeter measures expansion and contraction of crustal rock, indicating strain accumulation or release.

245

Laser-ranging instrument measures round-trip travel time of a light pulse and thus the precise distance between two points, indicating horizontal movement across a fault.

Tiltmeter reflects changes in land slope by the rising or falling water level in two containers.

Scintillation counter signals the amount of the radioactive gas radon released into groundwater by rocks under stress. Gauge shows changes in water pressure.

Seismometer detects even the slightest ground tremors. It reveals changes in seismic-wave velocity, a precursor of shocks.

Creepmeter is essentially a wire stretched across a fault to register any horizontal movement.

246

In the desert east of Los Angeles, a group of rodents has joined the quest for knowledge. Biologist Robert Lindberg and geophysicist Durward D. Skiles, both from the University of California, have installed pocket mice and kangaroo rats in artificial burrows. Three earthquakes, with magnitudes from 4.8 to 5.2, have given the rodents a chance to show their stuff. Lindberg and Skiles wired the burrows to record the animals' activity level before earthquakes. Unlike most previous observations, which depended on anecdotes and human memory, the rats and mice are part of an experiment conducted under scientifically controlled conditions.

Unfortunately, the best laid plans of these men and their mice sometimes go awry. Before the series of three earthquakes, the animals did do a lot of scurrying about. It turned out, however, that the quakes had hit during mating season.

But the two scientists are not discouraged. They have put more rodents to work in a colony in Hollister. In Albuquerque, New Mexico, seismologist Ruth Simon has a computer watching some cockroaches. A group of chimpanzees being studied at Stanford University's School of Medicine just happened to live next to the San Andreas Fault. They showed measurable agitation before some 1975 quakes.

In 1979 geologist Donald Stierman was camping in the Mojave Desert. There was an earthquake with audible booms. Dogs barked. Then came small aftershocks, when Stierman watched his seismometer record jolt after jolt. He neither felt the shaking nor heard any booms. But the dogs barked every time.

Look in the Yellow Pages

Earthquake safety and prediction are worldwide concerns today, and science is making strenuous efforts to study the physical properties of tremors. The Haicheng quake came after a ten-year pattern of shocks that interested seismologists like Christopher H. Scholz at Lamont-Doherty Geological Observatory. Scholz suggests that pre-quake tremors as well as the big shocks may be set in motion by "deformation waves" slowly rising and falling in Earth's mantle, waves which now and then crest at a weak spot in the crust. Don L. Anderson and others at the California Institute of Technology demonstrate their version of this undulating-mantle theory with a rippling Slinky toy.

If you live in San Francisco, the telephone book will tell you how to prepare your home for an earthquake and what to do if one strikes. Government and private agencies in many parts of the country work at the tricky job of earthquake forecasting. The newest and most delicate instruments are only part of the story. People are the other part.

Earthquake prediction is a sure thing along a "seismic gap," a normally active stretch of fault which has been quiet for a long time. The prediction: Someday it will slip.

But to be useful in densely populated areas of the world, earthquake forecasts have to be much more precise than that. Those who study the science of prediction know their work is going to take a long time. These scientists worry about crying wolf or raising panic. Yet, to protect lives and property, people still need to know *when, where,* and *how big.* These are tough questions.

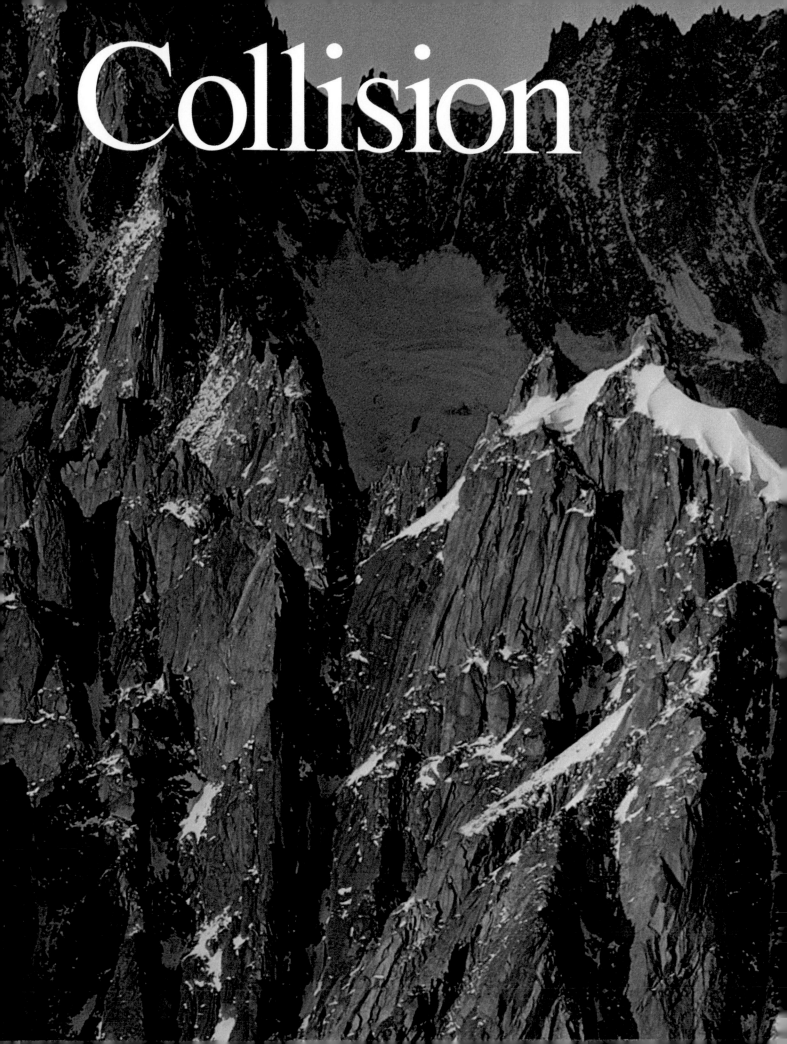

Collision

OVERLEAF: *Mont Blanc's rocky talons, thrust aloft as Italy presses into Europe, rake the French skies. Today's Alps began to rise only three to four million years ago.*

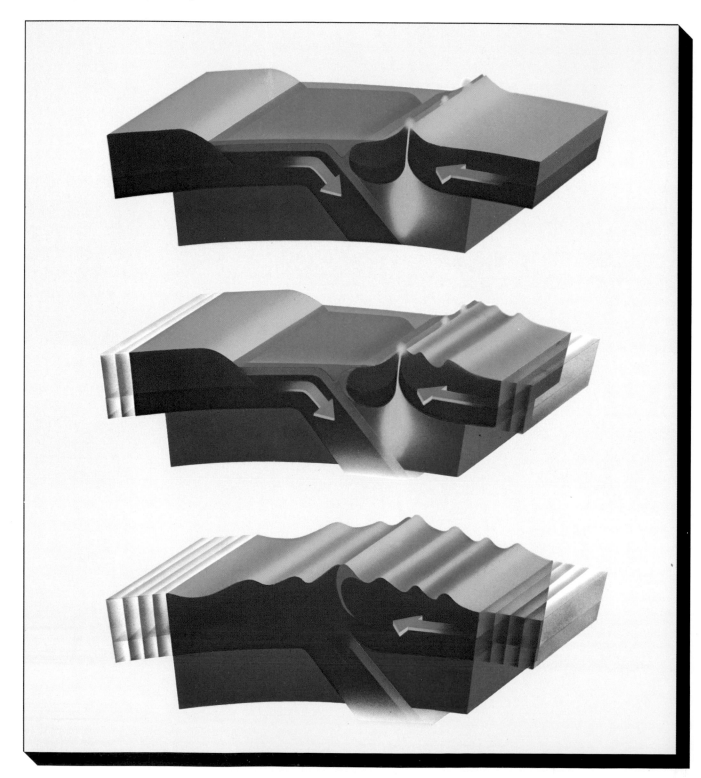

Collision:
Mountain Maker

The slow-motion crash *of two continents impends as converging plates close an ocean. The ocean floor, made of dense basaltic rock, plunges under one of the continents into a subduction zone marked by a deep trench (opposite, top). As the seafloor dives, it releases water, causing the overlying rock to melt. This magma rises, some of it all the way to the surface, where eruptions build a line of volcanoes parallel to the trench.*

The steady impact of plates causes one of the landmasses to crumple into parallel mountain folds (center). Often the leading edge of the continent's plate scrapes deep-water sediments off the diving plate, building up an accretionary wedge. A section of the diving crust may also break away, escaping subduction. As the rest of the seafloor disappears into the mantle, the continents meet head-on. Melding together at a suture, *they can pinch any leftover oceanic fragments up above water, even into high ridges (bottom). Bits of ocean crust stranded on land in this way are called* ophiolites. *Geologists today find sedimentary rock and ophiolites in the Alps, Himalayas, Appalachians, and other ranges far from the sea.*

Around the world, mountain ranges and ocean trenches identify collision zones in blue (map).

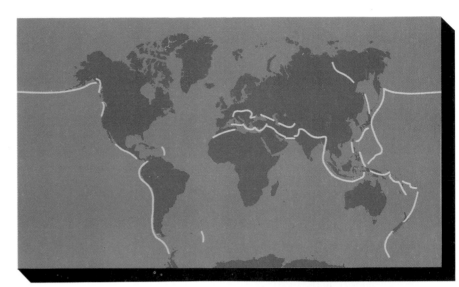

Curved island arcs *characterized by fiery volcanic chains (right) rise where two plates bearing ocean floor collide. Earth's spherical form forces the line of subduction to bend, shaping the distinctive arc. As the subducting plate dives into the mantle, the action pulls and stretches the overriding plate, causing the crust to crack. Molten rock in the mantle rises behind the island arc and creates a small spreading zone, a miniature version of the Mid-Ocean Ridge. This phenomenon, called* back-arc spreading, *formed the floors of the Sea of Japan and the Philippine Sea.*

*at subduction zones, volatile
magma rises and breaks
through Earth's crust, creating
volcanoes all around the
Pacific Ocean—the Ring of Fire.*

Earth must go back to Earth.
Cicero

April 1, 1946: Francis P. Shepard, a distinguished marine geologist from the Scripps Institution of Oceanography, and his wife, Mary, were living at Kawela Bay on the north coast of Oahu, Hawaii.

At 2:30 a.m., an earthquake measuring 7.5 on the Richter scale rocked the ocean floor in the Aleutian Trench 3,500 kilometers (2,175 mi) to the north of them. Like a great paddle, the heaving seabed pushed at the water. The first wave radiated outward from the epicenter at 780 kilometers (485 mi) per hour. Ships in the North Pacific did not notice its passing, for unlike a normal wave, which concentrates its energy at the surface of the sea, a seismic sea wave spreads its energy from top to bottom. In deep ocean, a seismic wave—called a tsunami, Japanese for "large harbor wave" —raises a gentle swell no more than a few feet high.

About four hours after the Aleutian quake, the tsunami approached Hawaii. The shallower seafloor close to land compressed the wave and focused its power.

"We were sleeping peacefully," said Shepard, "when we were awakened by a loud hissing . . . which sounded for all the world as if dozens of locomotives were blowing off steam directly outside our house." Startled, Shepard rushed to the window and looked out toward the beach in front of their small house. "Boiling water was sweeping over the ten-foot top of the beach ridge and coming directly at the house." Forgetting his clothes and glasses, he rushed for his camera but returned too late. The water was retreating—and it kept on retreating until it exposed a normally submerged coral reef, where stranded fish now flapped helplessly. Realizing he had just witnessed a tsunami, the so-called tidal wave, he warned his wife, "There will be another. . . ."

In 1923, the people of Hilo, Hawaii, had seen a tsunami retreat and were tragically fooled. The water receded *before* the arrival of the first wave. When they innocently ran onto the exposed seafloor to pick up the stranded fish, they died beneath the oncoming crest.

Now Shepard watched as a wave again began to build up outside the reef. "It built higher and higher and then came racing forward with amazing velocity." He took pictures until he realized this wave was much bigger than the previous one. He turned and ran to the back of the house as water rushed in. "Suddenly we heard the terrible smashing of glass. . . . The refrigerator passed us on the left side, moving upright out into the cane field." The wave carried a nearby house several feet inland and gently deposited it in the same cane field.

Shepard and his wife now headed for high ground. They literally ran for their lives as another great wave began to build. "Rising as a monstrous wall of water, it swept on after us, flattening the cane field with a terrifying sound. We reached the comparative safety of the elevated road just ahead of the wave."

Three more waves roared in as they watched. Finally, after the sixth wave, each separated by about 15 minutes, the worst seemed over. Shepard returned to the cottage to check the damage. Just as he reached the door, he looked up. He had made a terrible mistake. The biggest wave of all was heading toward him. With nowhere to run, he climbed a tree. The wave hit, swaying him back and forth. The Shepards lived, but 159 others died.

After the 1946 tsunami, authorities established a formal warning system. Tied to a network of seismological observatories and tide stations, the system issues watches and warnings around the Pacific. In some locales the warning may be only hours ahead of the tsunami, so local police and civil defense workers must move quickly.

But warnings present problems. If authorities give too many—they are based in part on difficult-to-measure undersea earthquakes— and no large waves follow, the population shrugs the warnings off. A less obvious danger is that curious people will flock to the beach. After a tsunami warning prompted by the great Alaska earthquake in 1964, some 10,000 people appeared on the beaches of San Francisco to wait for the wave to arrive. Fortunately, it never did.

In Crescent City to the north, however, the tsunami caused the greatest damage on the California coast in the last one hundred years. Just as Francis Shepard had done 18 years earlier in Hawaii, many Crescent City citizens returned to clean up after the first two minor waves hit, only to be caught by the third and fourth waves. The owners of the Long Branch Tavern had returned with five others to rescue money from the restaurant. Everything appeared normal, so they

Concentric arcs record the life of a killer tsunami that raced across the Pacific Ocean in March 1964. Born of a quake near Valdez, Alaska, this seismic sea wave fanned out at speeds of more than 800 kmph (500 mph), killing 122 people as it slammed into shorelines around the Pacific. The toll might have gone higher but for an oceanwide tsunami warning system based in Hawaii, and since then expanded to Alaska. The system's operators issue alerts based on tide gauges and earthquake detection.

In subduction zones circling the Pacific Basin, colliding plates spawn such quakes, as well as the volcanoes (gold dots) that inspire the name Ring of Fire.

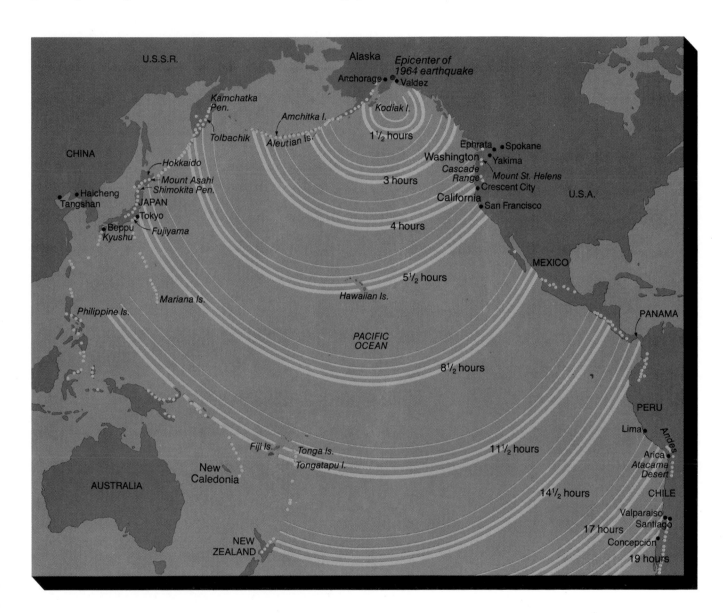

"We . . . stared, unable to believe what we saw"

Disbelief repeatedly gripped a Lieutenant Billings, eyewitness to terror the afternoon of August 13, 1868. From the side-wheeler U.S.S. Wateree, *anchored in the harbor of Arica, Peru (now part of Chile), he helplessly watched a nightmare unfold as an earthquake leveled the town. An hour later the first wave of a tsunami slithered under his vessel and engulfed frantic survivors clustered on the wharf. Then the waters receded until the harbor floor lay exposed, leaving other ships lying on their beam-ends. But the flat-bottomed* Wateree *remained upright. When the sea returned, it smashed one ship against the promontory called El Morro. In an eerie aftermath, Billings saw a spooky tableau on the mountain slope: The quake had opened tier after tier of tombs— and there sat the mummies, still wrapped in cloth and squatting in the burial position.*

OVERLEAF: *The sun had set when the* Wateree's *lookout spotted the next wave. "We made out a thin phosphorescent line . . . rising higher and higher." Below it loomed "frightful masses of black water." Nearly 15 m (50 ft) tall, the wave swallowed the* Wateree *for a "suffocating eternity" before depositing her 3 km inland, high and dry once more—just 61 m short of destruction at the foot of El Morro. Only one man was lost. Billings called it "a miracle."*

Glaciated peaks of the Andes seem to float like islands in the sky north of Lima, Peru. As oceanic crust of the Nazca Plate slides under continental crust of the South American Plate, volcanism and compression thrust up this mighty coastal mountain chain.

stayed for a beer. The third wave surprised and killed them.

That tsunami was the last major one to surge across the Pacific, but tsunamis will continue to endanger Pacific coastlines. For the earthquakes that cause them are but one manifestation of a continuing, far greater force: subduction.

The perimeter of the Pacific Basin is under attack. Every year the floor shrinks, eaten away at the edges by at least six overriding plates (and only partially replaced by new crust that grows out of the Mid-Ocean Ridge). Old ocean floor disappears into subduction zones at a rate that could devour the entire surface of Earth every 160 million years. The consumption of this crustal skin generates some of Earth's most violent internal actions. As the subducting plate descends, it grinds against the overriding plate, setting off the earthquakes that launch tsunamis pulsing across the Pacific. It also causes volcanic eruptions, mountain building, and even some seafloor spreading in back-arc basins.

This zone of violence circles more than 48,000 kilometers around the Pacific—the Ring of Fire, it is called. Here 75 percent of Earth's active land volcanoes tower above the offshore deep-sea trenches that mark the lines of collision.

Subduction zones vary. At some, a continental plate collides with an oceanic plate. At others, two oceanic plates meet. Some plates meet head-on, while others make only glancing contact. Earthquakes hit some zones often and hard, but leave others less shaken. Only in the last few years have scientists made any sense of the diverse aspects of subduction zones. Let's try to understand them by touring the Ring of Fire.

Along the Peru-Chile Trench, the Nazca Plate—an oceanic plate—plunges beneath the continental crust of the South American Plate. East of the trench the collision creates a parallel 8,850-kilometer (5,500-mi) mountain range, the Andes. They stretch unbroken from fiordlike Tierra del Fuego to the jungles of Colombia and Venezuela, shaping the region's special climates, habitats, life-styles, and distinct human physiology.

In Peru and northern Chile, the Andes force aloft moisture-soaked clouds spawned by the South Atlantic until the clouds cool, condense, and spill rain upon the eastern slopes, as much as four and a half meters annually. With the air squeezed nearly dry, little moisture remains to fall on the western slopes. In this rain shadow the coastal desert is lucky to get a few centimeters of moisture a year, usually condensed sea fog from the cold Peru Current.

Ancient Chimu lose to tectonics
Centuries ago on the coast, a people called the Chimu conquered the desert—until tectonic movement overpowered them. One of the great gold-rich pre-Inca cultures, the Chimu flourished from A.D. 600 into the 1400s. They built a sophisticated network of canals along the coast north of Lima, Peru. Rivers flowing from the mountains fed channels that irrigated a third more land than present-day systems in the same area.

But slowly, decade after decade, century after century, the system

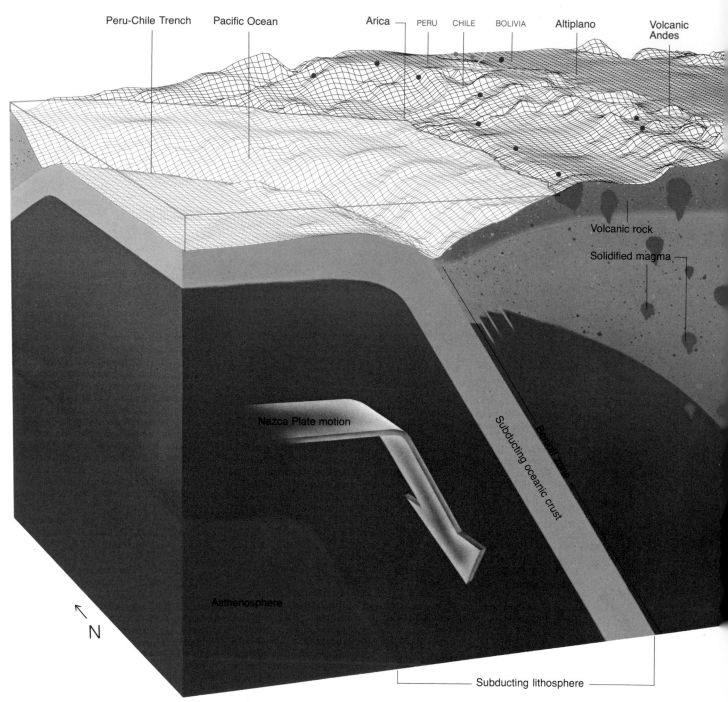

Peru-Chile Trench   Pacific Ocean   Arica   PERU   CHILE   BOLIVIA   Altiplano   Volcanic Andes

Volcanic rock

Solidified magma

Nazca Plate motion

Subducting oceanic crust

Benioff zone

Asthenosphere

N

Subducting lithosphere

The Andes' Infernal Forge

From miles down in the mantle, the valuable metals of the central Andes ride upward in pods of magma. Some lodge in rock near the surface, where erosion eventually puts them within reach of man. They form strands of wealth—copper, tin, lead, zinc, and silver—each running parallel to the offshore trench where the Nazca Plate begins to deflect

under the South American Plate.

Geologists focus their search for ores within these well-defined belts. They suspect the metals are carried upward in two ways—in hot water contained in the rising magma, and in groundwater heated by the magma. In both cases, the metals precipitate when the water cools. Deposits of the first type—such as porphyry

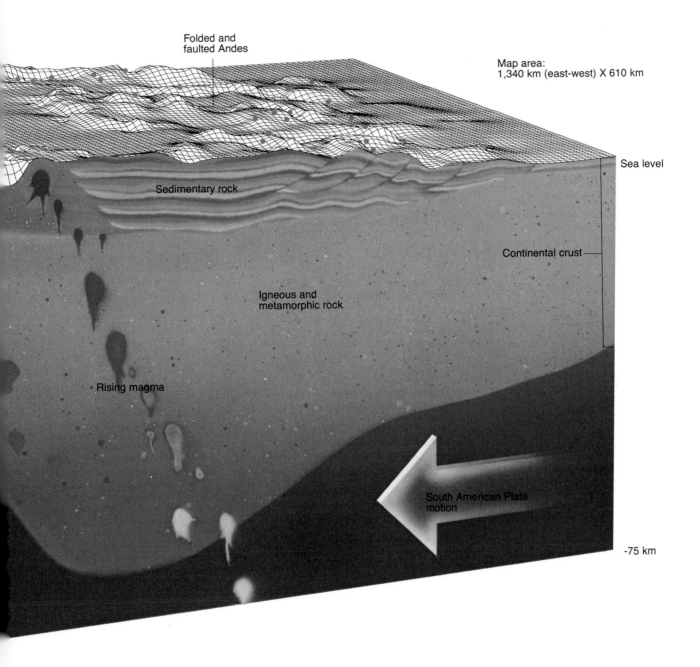

Folded and
faulted Andes

Map area:
1,340 km (east-west) X 610 km

Sea level

Sedimentary rock

Continental crust

Igneous and
metamorphic rock

Rising magma

South American Plate
motion

-75 km

copper—are large and lie about
70 km apart. The second type,
usually containing metals such as
lead, zinc, and silver, are much
smaller and closer together.

Folding and faulting raised the
Andean peaks east of the
Altiplano, or "high plain." The
newer, western range arose from
volcanic eruptions as magma pods
reached the surface still hot.

Mining centers

● Copper

● Tin

● Lead, zinc, and silver

began to fail. The coastal region rose, pushing the river basins up a couple of centimeters a year. The rivers cut deeper into the rising ground, eventually dropping below the mouths of the canal system. Over a thousand-year period, the growing inability of water to enter the irrigation system probably caused a loss of about a fourth of the arable land.

By 1476 the Chimu people were ripe for conquest by the Incas, who took Chimu artisans and gold back to the city of Cuzco high in the Andes. As the highlanders conquered other peoples and tribes, Inca engineers built a road system surpassing that of imperial Rome. Along 16,000 kilometers of cloud-hung highways, from Colombia to Chile, they transported grain, textiles, wool, jewels, and artifacts of gold and silver mined from the ore-strewn land.

Altitude shapes bodies
The cold, thin air of the high altitudes forges physically distinct people in the Andes, many of them descendants of the Incas. Short in stature, they have large upper torsos with unusually large lungs. Heart rates are slow. To help store oxygen—at 3,600 meters (11,800 ft) there is only 40 percent of the oxygen available at sea level—their bone marrow produces more red blood corpuscles than is normal for lowlanders. Blood pressure is low and rate of flow increases to arms and legs in cold weather.

Unfortunately, such robust bodies couldn't save the 16th-century Incas from European intruders. In the end the subduction zone beneath the mountains betrayed the Indians, for it gave their lands fabu-

lous mineral treasure—and treasure attracts treasure hunters. With a force of less than 200 men, Francisco Pizarro in 1532 conquered the Inca empire and melted down a king's ransom in gold and silver artifacts for shipment back to Spain.

The Spanish thirst for precious metals was unquenchable, and in the centuries after Pizarro hundreds of prospectors and miners scoured the Andes. Their greatest find was an old volcano in what is now Bolivia—Cerro Rico, the "hill of riches," which stood nearly four kilometers above sea level. Its silver veins, three and a half meters thick, built the mining town of Potosí. At the beginning of the 17th century Potosí boasted 200,000 people, the largest population of any city in the Western Hemisphere.

Why were the Andes so rich in valuable metals?

In the early days, as miners delved deeper and deeper, they felt part of the answer in their sweat: heat. Far below the mines, geologists speculate, the following process takes place. The Nazca Plate, scraping under the overriding South American Plate, transports great amounts of seawater into the lithosphere. At a depth of 100 to 200 kilometers, the water reduces the melting temperature of the rock, which liquefies into magma. Rich concentrations of copper and other metals—perhaps borne down by the subducting plate, perhaps already present in the surrounding rock—sweat out of the rock and rise with the magma. Most of the magma cools and solidifies before breaching the surface. But millions of years of erosion can bring the surface—and the miner's pick—

"This mountain is a powder keg...."

David Johnston's observation proved tragically prophetic. Mount St. Helens blew up on May 18, 1980, in the most publicized volcanic eruption of the century. The blast interrupted Johnston as the government geologist tried to radio in from a monitoring station near the peak. He was never seen again.

Shaken by a 5.0-magnitude earthquake deep in its bowels, Mount St. Helens tore 390 m off its crown and spewed a cloud of smoke and ash 21 km into the stratosphere. Out of its north flank roared a holocaust of hot debris and scalding gases at 320 kmph (200 mph), triggering devastating mudflows and floods, and claiming the lives of 64 people.

OVERLEAF: *Menacing clouds of ash, three hours after the eruption, seem ready to consume Ephrata, Washington, 233 km (144 mi) northeast of Mount St. Helens.*

2ND OVERLEAF: *Choked by St. Helens ash, Yakima, Washington, 136 km (84 mi) to the east, digs out from under 600,000 tons of the gray grit. Here heavy equipment battles the half-inch ashfall along Route 1. As far east as Spokane, ash clouds closed schools, clogged car engines, and curtailed travel by lowering visibility.*

267

within reach of these valuable ores.

Nature extracts metals from the subduction zone in a methodical way far exceeding the dreams of ancient alchemists. As the descending plate angles down to hotter and hotter depths, one metal after another separates from the others: iron, copper, gold, lead, zinc, and silver. As a result, South America's western ore deposits rest in belts roughly parallel to the offshore trench.

Do smokers enrich the Andes?

Geologists have argued about the source of these metals for years. Some believe the ordinary magma that creates the floor of the Nazca Plate does not contain enough metal to produce the concentrations of ores found in the Andes. Something, therefore, must have enriched the rock, and the answer may lie in the hot springs and black smokers we discovered on the Mid-Ocean Ridge. Metal compounds precipitate from the spring water and collect in the smoker chimneys, which crumble into the seabed and ride the Nazca Plate toward South America. Carried into the subduction zone millions of years later, these concentrated masses melt to form mineral riches.

Strung out through the Andes' mineral belts are the active volcanoes that give the Ring of Fire its name. Most lie about 100 to 400 kilometers inland from the trench in a systematic line, a volcano every 50 to 70 kilometers or so. They have been relatively quiet in modern times, sparing Andean peoples the fiery wrath others have suffered along the Ring of Fire. Yet the Andeans have had to endure centuries of devastating earthquakes.

Violent quakes in the Andes commonly occur at shallow depths, often near the offshore trench, as in the 1960 Chilean disaster. The reason has to do with the type of subduction zone.

In most subduction zones, the plates scrape past each other as far as 700 kilometers down—much deeper than the area of friction in translation zones like the San Andreas system. Earthquakes occur as long as the subducting plate and the rock it slides under resist one another. But eventually the descending plate becomes as hot as the surrounding material. At that point it finally reenters the mantle; friction—and earthquakes—cease.

The depth of individual earthquakes and their distances from the trench can provide a cross-sectional view of a subduction zone, indicating whether a plate enters at a shallow or steep angle of descent. The centers of these earthquakes fall on an inclined plane called the Benioff zone, after California Institute of Technology geophysicist Hugo Benioff, who plotted their positions in the 1940s and 1950s. Around the Ring of Fire, Benioff zones are associated with ocean trenches. Recognition that Benioff zones identify subducting plates was central to the early development of plate tectonics theory.

Along the Andes, where the Nazca Plate tilts down gradually, earthquakes regularly threaten dense population centers on the coast. In 1822, some 10,000 people died in Valparaiso, Chile. Concepción and Santiago were leveled in 1835. Valparaiso was hit again in 1906, the year of the San Francisco quake. Thirty years later a

Loggers tackle a massive cleanup in Mount St. Helens' blowdown area: 595 sq km (230 sq mi) embraced by Gifford Pinchot National Forest. They salvaged 600 million board feet of lumber, enough to build 60,000 three-bedroom houses.

St. Helens painted hundreds of ponds and puddles a riot of color. Minerals in her ash fallout favored the growth of different kinds of algae and microorganisms that tinted waters yellow, green, brown, blue, or a combination of hues.

274 three-minute convulsion in Concepción killed 50,000 people and left 750,000 homeless.

Such catastrophic losses may be greatly reduced if science can perfect ways to predict quakes—as the Chinese seemed to do in the winter of 1974-75 when their careful observations saved tens of thousands of lives around Haicheng. But the Chinese methods are not foolproof. A year after the Haicheng success, more than 650,000 people died at Tangshan, when a quake struck entirely without warning.

Until forecasts become routinely accurate, predictions could backfire. In 1981 the prediction of a killer quake for Peru jammed panicked people onto flights outbound from Lima. Those who stayed behind hoarded food, water, and medicine; relief agencies stockpiled body bags. When the day of destiny arrived—June 28—nothing at all happened. In the meantime, Peru's tourist industry suffered.

Such false alarms are worrisome. But until earthquake prediction becomes more exact, scientists may face hard decisions: Should they predict a quake, if the signs are unclear? If the quake does not come, the forecast could cause needless panic, as in Lima. But what if they withhold the prediction, and the quake does come, killing thousands who could have been warned?

A hurricane-force blast
The North American coastline does not have a continuous mountain range like the Andes. And for most of its length no major ocean trench identifies a subduction zone, because the North American Plate and the Pacific Plate are not collid- ing head-on but sideswiping each other along the San Andreas Fault. Nonetheless, North America does experience active subduction farther north.

Mount St. Helens in Washington State provided the Ring of Fire's extravaganza for 1980. According to a Klamath Indian legend, a great war of the gods, between the Chiefs of the Below World and the Above World, created Mount St. Helens and other volcanic peaks of the Cascade Range.

For the last 120 years, snow-capped St. Helens slumbered. But in March 1980, seismic war drums sounded a reminder of the ongoing battle: The small Juan de Fuca Plate continued to thrust beneath the North American Plate, and rising magma, sign of the tectonic struggle, made the ground tremble.

The Juan de Fuca Plate is a relic of the Pacific's past. Geologists believe that a huge plate they call the Farallon once formed the floor of the eastern Pacific. Now the Farallon has all but disappeared beneath the overriding North American Plate, swallowed by the Ring of Fire. Only two important fragments remain: the Juan de Fuca Plate and the Cocos Plate off Mexico.

Late in 1980, I joined other earth scientists at the annual meeting of the American Geophysical Union. There I learned details of the eruption from experts who had been watching St. Helens. Early in the year, the north face of the mountain had begun to swell. Day by day it grew. Then, on May 18 at 8:32 a.m., an earthquake shook the mountain and a giant avalanche of debris thundered down the north slope. Subterranean pressure was suddenly

276   released. Water circulating in and near the magma chamber beneath the mountain flashed into steam and a great explosion blew out the north side of the volcano. With the energy of a ten-megaton nuclear bomb, a *nuée ardente,* or glowing cloud, of ash, gases, and pulverized rock raced down the mountain.

At a U. S. Geological Survey station 9.6 kilometers (6 mi) from the mountaintop, young David Johnston tried to radio in his report. He got as far as "Vancouver! Vancouver! This is it—" before he disappeared in the 200-mile-per-hour hurricane of scalding gases and fire-hot debris. Harry Truman, the 84-year-old lodge-keeper who had lived for half a century beside Spirit Lake beneath the peak, refused to leave his beloved mountain. It buried him under hundreds of feet of ash, debris, and water.

In the aftermath, plants and animals have returned. Perhaps the most bizarre residents are chemical-eating bacteria similar to those we found feeding on the mineral soup emerging at the Mid-Ocean Ridge. For about a year they lived in Spirit Lake, possibly carried by the bombardment of ash. Though they disappear as the lake cleanses itself, they still exist in the water inside the crater's dome. How they got there remains a mystery, though some scientists believe they could have come from an underground hydrothermal plumbing system.

Mount St. Helens is one of the most thoroughly instrumented volcanoes in the world. Its earthquake activity is carefully monitored, and tiltmeters measure the slightest deformation of the ground as the subsurface magma chamber expands and contracts like a beating heart.

Although no deep trench reveals the edge of the subduction zone that feeds this chamber, Mount St. Helens stands about 190 miles inland from the line where the Juan de Fuca Plate disappears. This distance between ocean plate and rising magma is common to most subduction zones, a sign that the plate causes melting only after it has subducted some distance.

Subduction zones may also be marked by the curved shape of the volcanic chains above them. When formed at sea, these chains are called island arcs.

Bombing the basement of the Aleutian Islands

At the northern rim of the Pacific Plate, the Aleutian Islands stretch out in a 735-kilometer-long (456 mi) arc. Just south of this arc runs the gently curved Aleutian Trench. The curve comes from Earth's round shape. If the surface of a sphere is cut into at an angle (by a subducting plate), the cut (the trench) will assume a curved shape.

The Aleutians lie over an area of particularly fragile crust. Some 50 million years ago, when the Farallon Plate occupied the eastern Pacific Basin, the triangular Kula Plate formed the floor of the northern Pacific. Today, the Pacific Plate has pushed the Kula Plate entirely into the Aleutian Trench and is diving in after it. Pressure from subduction has left the northern rim of the plate badly frayed and faulted and thus prone to quakes.

When the Atomic Energy Commission (AEC) decided to test a nuclear bomb a mile under the Aleutians, it raised a storm of

Fiery tongue of Tolbachik volcano licks the sky over the U.S.S.R.'s Kamchatka Peninsula. Spewing gases and incandescent magma continuously for three weeks in July 1975, it allowed scientists time for thorough study. Tolbachik and 21 other active peaks make the peninsula one of the most restless regions on the Ring of Fire.

controversy. Underground nuclear testing had begun in the deserts of Nevada, but large bombs swayed buildings in nearby Las Vegas. Seeking a remote site, the AEC in 1966 chose Amchitka Island in the western Aleutians. Many people feared that the explosion, calculated to be 250 times greater than the Hiroshima bomb, would trigger a quake like the one that struck Alaska in 1964. That quake was 4,000 times more powerful than the largest nuclear bomb ever tested.

On November 6, 1971, technicians detonated the bomb. For a fraction of a second, the temperature in its chamber reached almost two million degrees Celsius. A man-made earthquake measuring 7.0 rocked the island. The blast excavated a cavern estimated to be 60 meters (200 ft) wide, which collapsed 38 hours later, setting off another tremor of 4.8. Within the next three months, 22 earthquakes occurred that scientists think were triggered by the test. None, however, exceeded 3.5.

Using seismic information from the earthquakes and from nuclear tests, geologists drew a detailed picture of the Aleutian subduction zone. They discovered there, as they have in other places around the Ring of Fire, several "seismic gaps," areas where earthquakes should occur but haven't in the last 30 years. Scientists suspect that major strain, unrelieved by the small quakes often felt along the Ring of Fire, will eventually cause devastating quakes to strike these regions. One such major quake hit southeast Alaska in 1972, registering 7.6 at Sitka.

Seismic gaps also occur in Japan, one of many island arcs along the western part of the Ring of Fire. Japan sits where the Pacific and Philippine Plates dive under the Eurasian Plate. Japan's arc, like the Aleutian Islands, is segmented, due to breaks and rips in the subducting oceanic plates.

Unlike the Aleutians, however, Japan has a back-arc spreading basin between it and the relatively stationary continent of Asia. The basin opened when the subducting plates created suction that pulled Japan away from China. This caused the crust between China and Japan to stretch and fracture, allowing molten rock to well up and form new seafloor. Now inactive, the fractured zone beneath the Sea of Japan continued to create new seafloor until about 20 million years ago.

Ring of Fire promotes communal bathing

Heat from the subducting plates also feeds more than 40 active volcanoes and some 20,000 hot springs in Japan. These geologic features combine with religion to shape one of the country's enduring customs: communal bathing. Buddhism asks its adherents to purify themselves in clean water before praying. As far back as the eighth century, Japanese temples provided bathing facilities. Hot springs, particularly in winter, were favorite bathing sites.

In time, the priests noticed that continued bathing in hot springs had "miraculous godlike effects upon human sicknesses." Soon hot-spring sites had medical temples dedicated to the Buddha of Medicine. Today, there are about 1,800 hot springs of recognized medicinal value in Japan. It is said that sulfur springs are good for skin and liver problems; iron carbonate springs for the blood, nerves, mental fatigue, and female diseases; salt springs for diseases of the bones.

The Japanese use their subterranean energy in numerous ways. Hot-water-heated greenhouses nurture orchids year round. Research centers raise tropical fruit trees, cacti, exotic plants, and 20 different kinds of alligators and crocodiles. Hot-spring heat keeps chicken coops warm when cold winds blow in from the north. The heated floors of the coops also dry the droppings, keeping the chickens clean and the air breathable.

In the city of Beppu, an engineer has designed what may be the world's only hotel not dependent on fossil fuels. Under the Suginoi Hotel, a 1.3-meter pipe sinks almost 200 meters to an underground supply of hot water. Pumped up to heat exchangers, it supplies central heating and air conditioning. Geothermal power drives a 3,500-kilowatt generator for electricity. Geothermal greenhouses provide dining tables with vegetables. A heated aviary houses peacocks, and scores of bathers enjoy the hotel's two enormous hot tubs—each the size of a large swimming pool. Pipes even run beneath the lawn to keep the grass green in winter.

Nationally, geothermal power has become increasingly important to this industrialized people isolated on relatively young volcanic islands with a very small domestic oil supply. But the Japanese pay a price for their geothermal energy, a price common above subduction zones: earth tremors, 10,000 a year. Of all the seismic energy released in

Baking and Broiling in Japan

Sand bathers on the beach soak up heat from seaside mineral springs that make Beppu, on the island of Kyushu, the hot-spring capital of Japan. Some 4,000 springs began to flow after Mount Tsurumi, now sleeping above the town, erupted in the ninth century and created a hydrothermal field 243 m (800 ft) underground. Jigoku, or boiling ponds, provide instant mud baths.

281

OVERLEAF: Smoke and steam from sulfur vents turn Mount Asahi into an eerie wonderland for cross-country skiers on Hokkaido, Japan's northernmost island. A national park protects the 2,290-m (7,513-ft) peak, largest in the island's chain of volcanoes.

2ND OVERLEAF: Snow monkeys groom each other in a hot spring on the Shimokita Peninsula far north of Tokyo. Though monkeys originated in warmer climes, these Japanese macaques adapted as their climate turned frigid. Now they endure months of cold and snow, alleviated only by baths like this one heated by the Ring of Fire. Japan owes such thermal springs to its location above the subducting margins of the Pacific and Philippine Plates.

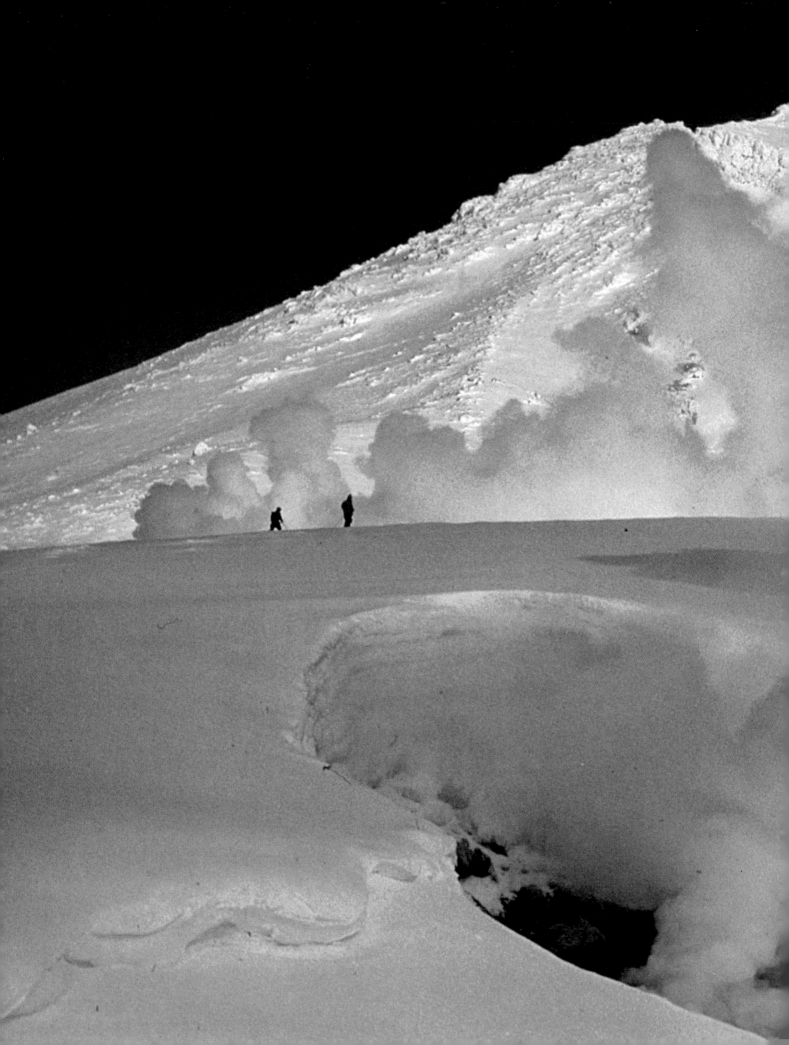

the world, 15 percent comes from the subduction zone beneath Japan. In the last 50 years, nine major quakes have killed over 12,000 people. When a quake struck in 1923 near Yokohama and Tokyo, 140,000 people died.

In Tokyo, as in San Francisco in 1906, fire did most of the damage. Hibachis filled with red-hot charcoal kindled paper walls into a fire storm. Winds from an offshore typhoon fanned the fires. Swirling sparks ignited the sails of ships anchored in Tokyo Bay.

The Japanese have been bracing for another major quake in the heavily industrialized Tokai district southwest of Tokyo. There, scientists have identified a seismic gap: No major earthquake has hit in the last 50 years. Moreover, in the last 85 years, the region has subsided 30 centimeters (12 in) as the subducting plates try to drag the east coast of Japan down with them into the Japan Trench. Earthquakes should be relieving the strain, but they aren't. Therefore, energy must be building for a violent vertical adjustment in what has already been named the "Great Tokai Earthquake." Experts say it might register 8.0 or more in magnitude.

Intense preparations are under way. Some 96 million gallons of water have been stored in "earthquake-proof" warehouses, enough for a ten-day supply for Tokyo's 12 million people. Disaster teams practice frequently and millions take part in earthquake drills. Some families assemble earthquake kits of fire extinguishers, flashlights, food, and water. September 1, anniversary of the 1923 quake, has become "disaster prevention day."

At a facility near Tokyo a seven-story building was erected so Japanese and American researchers could demolish it in an effort to learn how to build safer structures. On Japan's famous "bullet trains," which whistle along the rails at 209 kilometers per hour (130 mph), emergency brakes will deploy automatically when triggered by an earthquake shock wave of high enough intensity.

Japanese scientists hope, of course, to predict the time of the Tokai quake far enough in advance to evacuate people from dangerous sites. They feed a Tokyo computer with information from 70 different locations on crustal changes, surface tilt, magnetism, and gravity. They also monitor changes in land elevations above sea level, and the change in groundwater levels.

For a while, they even tried a method pulled from mythology. In the distant past, says Japanese folklore, a giant catfish lurked under the islands. When it moved, the earth quaked. In the late 1970s, scientists kept several catfish swimming in a tank where they were scrutinized for erratic behavior that might signal the onset of an earthquake. The results, however, seemed to bring down the catfish theory. The fish missed quakes 19 times out of 20.

Finding the answer

We have now traveled almost completely around the Ring of Fire and discovered a number of facts about plate tectonics: There are different kinds of collisions. All spawn earthquakes and volcanoes, but some produce mountain ranges, others island arcs, many with back-arc basins. But we still don't know why.

Seawater spouts from a blowhole in the Tongas, an island arc formed where one seafloor dives under another. Along the deep Tonga Trench east of the islands, the southwestern edge of the Pacific Plate plunges under an oceanic piece of the Indian-Australian Plate. The location makes the Tongas a zone of active island building. An eruption on December 10, 1967, created a new island, but within two months waves had worn it away into another of the Tongas' many shoals—each the top of a volcanic seamount raised by collision of tectonic plates.

The answer may rest in the last section of the Ring, in the series of island arcs running between Japan, the Philippines, and New Zealand. Here at sea, where a heavy plate dives under a less dense, lighter plate, island arcs such as the Bonin, Yap, and Marianas have formed—like the Aleutians—above namesake ocean trenches. Back-arc spreading, as once occurred in the Sea of Japan, may take place in the lighter plate, actively adding to the seafloor (see diagram, page 251).

This phenomenon is at work in the Marianas, whose trench has the greatest depth on Earth. When scientists descended into the Mariana Trench in 1960 and touched bottom at 10,915 meters below sea level, they had gone four times deeper than I dived that summer of 1974 on Project FAMOUS.

According to Japanese seismologist Seiya Uyeda, the great depth of the Mariana Trench is characteristic of one type of subduction zone. Our first stop on this trip, the Peru-Chile trench-mountain system, exemplifies the other type.

Uyeda and his co-workers make a distinction between zones that have back-arc spreading basins and those that do not. The latter, called "Chilean," have many more major earthquakes, show signs of compression in the rocks, and seem to dip toward the mantle at a shallow angle. The Mariana type of zone has a back-arc basin and dives steeply toward the mantle. The Chilean type releases nearly a hundred times more earthquake energy than the Mariana type.

Uyeda and his American colleague Hiroo Kanamori, of the California Institute of Technology, theorize that the Chilean type of zone requires a more violent head-on collision, with friction causing the earthquakes. In the Mariana type, they say, the plates meet with far less force. Indeed, it may be that the two plates move in the same general direction, with one gradually overtaking the other. One piece of oceanic crust in the Mariana type tends to be older and thus denser; it sinks easily beneath its younger and less compacted neighbor, dropping steeply into the asthenosphere with subsequently less friction and fewer major earthquakes.

Although they are not sure why, Uyeda and Kanamori say back-arc spreading develops at Mariana zones because the kind of rising magma needed for spreading—basalt—is formed only where two plates make gentle contact.

The two types of subduction are not mutually exclusive. One type may change to the other, or stop somewhere in between. For example, Japan is now above a Chilean type of subduction zone. It is earthquake prone and the floor of the Sea of Japan is not spreading. But the subduction zone under Japan could have been a Mariana type until just a few million years ago. That would explain how the Sea of Japan formed, and why it no longer actively opens.

No one knows precisely, of course, the hows and whys of subduction zones or the primordial violence attending them. But thanks to scientists like Uyeda and Kanamori, and to those dedicated bands of earthquake and volcano watchers—perhaps even to smarter catfish—we can continue to plumb the mysteries of the Ring of Fire.

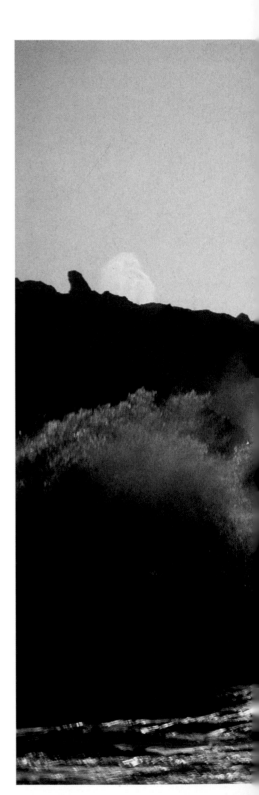

a seated Buddha contemplates Merapi (right), killer volcano that doomed the flourishing kingdom of Mataram in 1006.

How loud was the mightiest blast from Krakatoa? As the map below shows, the Indonesian volcano's climactic detonation in 1883 reverberated as far away as Perth, Australia. A comparable explosion from Mount St. Helens would have resounded throughout the United States. Nearly 18 cu km of rock and ash blew into the air, plunging nearby areas into blackness for as long as two days.

More than 400 volcanoes dot the 5,600-km (3,500-mi) Indonesian island arc, where the Indian-Australian Plate dives under the Eurasian Plate. The Java Trench (lower), three times deeper than the Grand Canyon, marks the boundary between the plates.

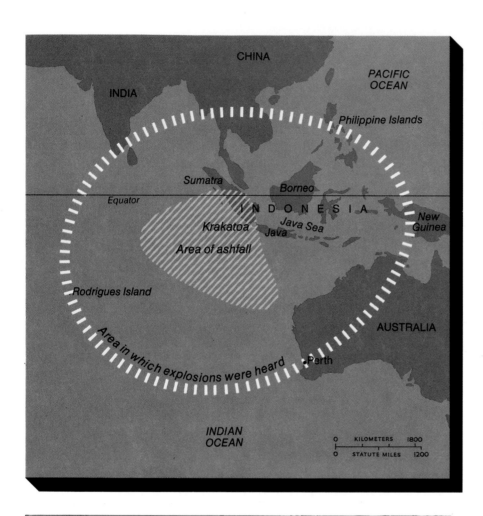

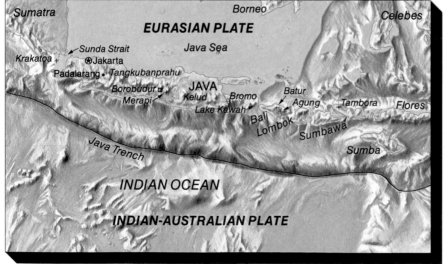

□ Archaeological site

Civilization exists by geological consent . . . subject to change without notice. *Will Durant*

On the island of Java in Indonesia, they tell the tale of an ancient volcano, Tangkubanprahu — "capsized canoe." Many years ago, according to the legend, there lived a prince who sought to marry his own mother. But the mother, a beautiful princess, feared the wrath of the gods over such a union and demanded of her son a seemingly impossible task: to create a lake in the mountains nearby and build a suitable nuptial canoe—all before the sun went down that day.

The prince set to with a will, damming a river gorge near the town of Padalarang and hewing a canoe from an enormous tree.

So the wedding took place. And the gods were not pleased. They sent down a holocaust of fire and water and thunder that overturned the canoe and drowned the participants. To this day, it is said, the upturned hull may be seen in the elongated shape of Tangkubanprahu.

Tales of volcanic upheaval such as this abound throughout Indonesia, one of several nations around the world with a history and geography that have been fundamentally shaped by volcanoes. In the mountains around Tangkubanprahu, scientists have traced the outlines of an ancient lake formed when landslides dammed a river. They have also discovered stone tools left by settlers who lived near the lakeshore around the time of Christ.

Indonesia's location above the subducting seafloor of the Indian-Australian Plate makes it intensely

Tossing a plume of smoke and ash into the air, Anak Krakatoa, "child of Krakatoa," rises from the ruins of its predecessor. In 1927, after lying quietly for 44 years, Krakatoa began to stir again, building a new cinder cone in the center of its submerged caldera. The cone grew intermittently, emerging from the sea during a period of intense activity in October 1952. Now more than 100 m high, Anak Krakatoa may someday follow in the footsteps of its parent and its grandparent, a mile-high volcano that erupted thousands of years ago.

volcanic, with about 100 active volcanoes in a land area a quarter the size of the United States.

The Sunda Strait, a major center of the archipelago's geological turmoil, lies between Java and Sumatra. It formed when fault blocks gradually sank, splitting one large island into the two smaller ones. In 1883, magma working its way up through those faults captured world attention by touching off a blast that demolished most of a volcanic island in the strait—Krakatoa.

Why are Indonesia's volcanoes so violent? When a dense oceanic plate collides with lighter continental crust—and much of Indonesia rises from flooded continental crust—the lighter, more buoyant plate overrides the heavier one. As my colleague Haraldur Sigurdsson explains, the descending oceanic slab then begins to melt, releasing water and gaseous liquids that slowly "bubble up" through the overlying plate. Each "bubble" melts its way through the continental crust, forming a "contaminated" magma charged with liquefied gases under tremendous pressure. Such magma is far less fluid than the lavas that ooze from spreading zones and many hotspots. As the bubble nears the surface, the gases within it begin to expand, fracturing the rock above and giving vent to an explosive volcanic outburst.

Java alone numbers more than a hundred volcanoes. A quarter of them have erupted within the last three centuries. Hot springs and belching fumaroles also attest to the island's continuing subterranean activity. The same conditions prevail on Sumatra, as well as on islands east of Java—Bali, Lombok,

Worshipers crowd the lip of Java's Mount Bromo at dawn on the feast day of Kesodo, an annual rite to propitiate the god of the volcano. They toss food, seeds, and flowers into the crater, where other villagers (below) catch the offerings—some to be used as seed crops for the next planting.

294 Timor, Flores. At least 16 volcanic cones cluster on Flores, which is not much bigger than Connecticut.

Small wonder, then, that earthquakes and volcanoes have had a profound, and sometimes devastating, effect on the people of Indonesia. One of the earliest known catastrophes involving humans, the eruption of Toba in northern Sumatra, took place 70,000 years ago. No one knows how many victims the explosion claimed. But tools and weapons belonging to prehistoric hunters have been found buried in ash across the Strait of Malacca on the Malay Peninsula—more than 400 kilometers away.

Mataram: solving the riddle of a vanished kingdom

The eruption of Tangkubanprahu probably brought on famine and a mass exodus from western Java to the central part of the island. The population shift, in turn, may have given rise to the fabled and ill-fated kingdom of Mataram, a cultural flowering that culminated with the construction of the wondrous temples and palaces of Dieng, Borobudur, Mendut, and Prambanan.

Then, suddenly, about A.D. 1000, the kingdom vanished from history and the great, half-finished temples of Prambanan were abandoned to the encroaching jungle.

For decades Western scholars puzzled over the mysterious fate of Mataram. Was it war? Was it pestilence? An inscribed stone, brought to Calcutta by a 19th-century colonial governor of Java (and hence known as the "Calcutta Stone"), proved maddeningly ambiguous. It described an idyllic empire where people lived happily, "like in the State of Indra." But then, without warning, they were overwhelmed by *mahapralaya*—utter chaos. The palace was destroyed; the monarch and his court perished. Only the king's son-in-law escaped, fleeing eastward over the mountains. The stone described events that had taken place during the year "928 Sjaaka" (A.D. 1006)—the year, subsequent investigation shows, that slumbering Merapi, "mountain of fire," blew its top.

Dutch geologist R. W. van Bemmelen examined the area around the volcano in the 1940s and concluded that the old kingdom of Mataram had been dealt two separate and distinct geological blows. The first silted in one of the few good natural harbors along the north coast, near the present town of Semarang. By around 927 the harbor had to be abandoned. Then, in 1006, Merapi exploded in a catastrophic eruption that buried central Java under a deep layer of ash.

All told, perhaps a quarter of a million people have been killed in Indonesia over the last century alone by volcanic eruptions and their aftereffects. Visitors to Java have commented on the sense of brooding menace that counterpoints the island's lush, hilly beauty. An English traveler saw in the 1919 eruption of Kelud "a striking example of those overwhelmingly sudden transitions from extreme peace to moods most terrible which characterize Java and lead one to mistrust her beauty."

As noted, one of the most widely known eruptions was the explosion of Krakatoa in the Sunda Strait during the summer of 1883. Krakatoa, a remote and uninhabited island,

Fertile Fields
From Bursts of Fire

Shimmering paddies garland volcanic slopes near Muncan village on Bali. Terrace walls retard erosion, helping to retain nutrients that make the island soil fertile. Surveying a three-week-old rice crop, a farmer balances a tool used to cut and repair terrace walls. Volcanic ash periodically enriches Balinese soil with many minerals, including phosphorus, calcium, and trace elements vital to plant growth.

297

OVERLEAF: *Batur, one of Bali's sacred volcanoes, spews a fountain of fire during eruptions in the summer of 1971. Dormant for four decades, Batur roared to life in 1963, forcing hundreds of villagers to flee. Balinese also revere neighboring Agung, their other holy mountain—which erupted three times that year.*

2ND OVERLEAF: *Harvesting sulfur along the shores of Lake Kawah in Java, a worker hefts a 50-kg (110-lb) load for the arduous trek to a mill 20 km away. The mineral, when processed and combined with phosphate, makes fertilizer. A vent in the crater of Kawah Ijen produces the sulfur, often amid clouds of noxious fumes.*

Weighing in after their journey to Kawah Ijen's crater, these men wait for their pay—about eight dollars for each load of sulfur.

Women on the Ayung River (right) scoop volcanic sand, an ingredient in cement used to construct Javanese roads.

rose some 800 meters above the strait, occupying part of a caldera left by an even more massive volcano that had exploded in prehistoric times. After Krakatoa's last recorded eruption in 1680, it had lapsed into dormancy for two centuries—just one more picturesque volcano lending exotic charm to the islands of the East Indies.

All that ended on the morning of May 20, 1883, when, with a roar that rattled doors and windows more than 160 kilometers away, Krakatoa exploded in a paroxysm of smoke, steam, and ash. And that was just the opening salvo.

Nights of thunder, days of darkness

For the next three months intermittent volcanic activity continued. Then, on the afternoon of August 26, a series of extraordinarily violent detonations began, collapsing part of the volcano's cone and touching off a succession of seismic sea waves that battered the shores of Java and Sumatra.

Finally, at 10 a.m. on August 27, the volcano erupted in a cataclysmic outburst heard in Perth, Australia, 3,600 kilometers (2,200 mi) away. Almost instantly a gigantic black column of steam and ash shot 80 kilometers (50 mi) into the sky, blotting out the sun in a gritty, choking pall that lingered for days.

What was left of Krakatoa's cone collapsed into the now empty magma chamber, leaving a caldera 7 kilometers wide and nearly 300 meters deep. Giant waves, one of them more than 30 meters high, washed ashore across the strait, wiping out coastal towns and claiming at least 36,000 lives. In the aftermath of the explosions windblown ash and dust circled the globe for several years, and the name Krakatoa became indelibly associated with stupendous volcanic explosions.

But living in the shadow of a volcano can have compensations. Java supports one of the world's highest population densities, 663 people per square kilometer—which works out to about a quarter of the population of the entire Southern Hemisphere.

The island's volcanoes make this possible, periodically replenishing the soil with ash and dust rich in minerals. Java's *sawahs*—irrigated rice fields—rank among the world's most productive croplands. Thus, on Java, farmers move *to* volcanoes, rather than away from them, considering the risk well worth the bounty of fertile soil.

Java's geology makes other important contributions to Indonesia's economy. Besides the agricultural riches and the gold and silver traded by the old kingdoms, iodine, manganese, phosphate, sulfur, and tin are mined on the island.

More important still are the oil and gas deposits found in a petroleum geologist's maze of small, mostly offshore fields along Java's northern coast. There, the pressure of the incoming Indian-Australian Plate has produced complex folding in the layers of sedimentary rock, creating myriad small traps in which oil, gas, and other hydrocarbons have accumulated.

In the old days, volcanoes were venerated with regular rituals and offerings—occasionally human sacrifices. Even though most of Indonesia has long since embraced the faith of Islam, some of the old customs linger, especially in isolated

304 areas. The Tenggerese of East Java, for example, still annually honor the god of Mount Bromo with a pilgrimage to the top of the mountain and colorful ceremonies that coincide with the full moon on the 14th day of the Kesodo (tenth) month.

Ilopango and the evolution of Maya civilization

Volcanoes also played a crucial role in the course of Maya civilization in Central America, according to a theory put forth by anthropologist Payson D. Sheets of the University of Colorado. By the third century A.D., the Maya had developed a relatively sophisticated society in the highlands of what is now El Salvador. A major route along the Pacific coast enabled them to trade extensively with neighbors to the north and west—in commodities such as obsidian mined from nearby volcanoes, as well as jade, cotton, cacao, and salt.

But these highlands are also part of the Pacific's Ring of Fire, and sometime around A.D. 300, disaster struck without warning. In two massive and nearly simultaneous explosions, the volcano Ilopango blew up with such ferocity that the land around it for 100 kilometers (60 mi) was rendered uninhabitable. As much as 40 cubic kilometers of glowing ash and debris shot into the air, hurling trees like matchsticks and killing thousands of people.

Near the volcano, entire forests were uprooted and instantly carbonized. Dust and ash some 50 meters deep inundated the land. At Chalchuapa, a thriving town about 80 kilometers away, dust and ash at least half a meter thick covered buildings, forests, and fields.

With their crops and farmlands devastated, survivors fled to the lowlands of present-day northern Guatemala and Belize. The once bustling Pacific trade route, now unused, soon reverted to trackless rain forest. So ended Maya civilization in the highlands of El Salvador.

But the Ilopango disaster proved a boon to the lowlands. Tikal, formerly an out-of-the-way settlement in northern Guatemala, burgeoned into one of the chief centers of Maya civilization. Trade was rerouted through Tikal, and new prosperity brought skills and knowledge that helped the city rise to eminence. At the height of its power, between A.D. 300 and 900, Tikal grew to more than 50,000 people and covered an area of some 130 square kilometers. Thus the volcano may have contributed to the advancement of what scholars consider the Classic Maya period.

1816: the summer that never was

Dramatic and destructive as a volcano can be locally, a major eruption may also have lingering effects around the globe.

Dust and ash pumped into the atmosphere can create brilliant sunsets by diffracting, or bending, light rays as the sun nears the horizon. Such was the case with Krakatoa and the eruption of Alaska's Mount Katmai in 1912. Sunsets were so spectacular after Krakatoa's blow-up that, on one occasion, alarmed residents of Poughkeepsie, New York, summoned their fire fighting company to do battle with a nonexistent blaze.

Atmospheric sulfuric acid produced after the eruption of Lakagígar in Iceland, cooled the world's

Centers like Palenque flourished in the northern lowlands after Ilopango's third-century explosion removed competition from the southeast (right). Traders had to abandon their Pacific coast route and blaze a new one through Tikal.

Generator of weather—sometimes on a global scale—a volcano spews dust, ash, and sulfur dioxide, which turns to sulfuric acid in the atmosphere. Heavier particles settle out rapidly or are "scrubbed" from the air by rain. But a powerful explosion can hurl gas and dust into the stratosphere —beyond reach of cleansing rains.

Jet streams disperse the ejected material, drawing a veil across the globe. Dust reflects some sunlight, and sulfuric acid absorbs some more before it can reach Earth's surface. The haze eventually dissipates, but before it does, it may noticeably lower surface temperatures. Tambora's 1815 eruption brought summer frost and

snow to much of the Northern Hemisphere, making 1816 the "year without a summer."

Lightning jabs at the crest of Guatemala's Volcán de Fuego during an eruption in 1974 (lower). Static electricity, created by colliding dust particles, makes such pyrotechnic displays common during eruptions.

weather in the 18th century. And the explosion of Tambora on the Indonesian island of Sumbawa during the spring of 1815 may even have speeded the settlement of America's midwestern states.

In this eruption, considered the greatest in modern history, the uppermost 1,300 meters of Tambora blew away in a detonation that tore a six-kilometer-wide crater in the peak and killed 12,000 people. The blast, substantially more powerful than the one that shook Mount St. Helens, hurled about a hundred times as much rock and ash into the air. Caught by the jet stream in the upper atmosphere, far above normal weather altitudes, Tambora's massive cloud slowly circled the globe and, by the following summer, a volcanic dust veil blanketed most of the Northern Hemisphere.

As the summer of 1816 unfolded, a grim picture began to emerge. A succession of unseasonable cold waves swept through Canada, the eastern United States, and much of Europe, devastating all but the hardiest crops. It was the summer that became enshrined in American folklore as "Eighteen Hundred and Froze to Death."

On many days thermometers in New England barely reached 50°F, and at night temperatures often dipped below freezing. A blizzard on June 6 dumped up to six inches of snow on parts of New England and, on the Fourth of July, Savannah, Georgia, registered a high of only 46°F.

Frost, snow, sleet, and ice ruined crops in much of the Northeast. Famine stalked Europe and food riots broke out in parts of France. People ate everything from moss to

Lilies of a vanished civilization bloom again after a 3,500-year entombment in the ruins of Akrotiri, a once thriving town on the Aegean island of Santorini —part of a volcanic island arc. Today a five-island cluster, Santorini is all that remains of a single volcano that erupted catastrophically near the time of the mysterious end of Minoan civilization on Crete, around 1450 B.C. The explosion buried Akrotiri under 60 m of ash and pumice. Debris rained down on Turkey and on Minoan settlements such as Zakros on Crete. Known as seafaring traders, the Minoans might have lost part of their fleet to blast and accompanying tsunamis.

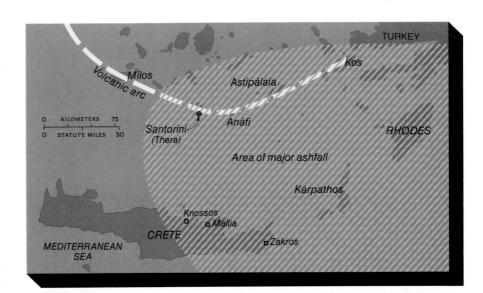

cats. Churches issued warning bulletins to help parishioners identify poisonous plants. It was the year, too, that recorded an exodus of Yankee farmers from New England to the greener pastures of Ohio, Illinois, and Indiana.

Tambora's effects were widespread, but some scholars speculate that a Mediterranean eruption in ancient times may have had an even greater impact on history.

Santorini and the Minoan mystery

From the cliff top I have a bird's-eye view of an Aegean cruise ship at rest below me, protected in the wide bay that lies inside the arms of this crescent-shaped island called Santorini. From the base of the cliff a switchback path of 587 steps ascends to the village at the top. You can ride up the path on a mule if you wish, and a band of tourists from one of the ships is doing just that.

I speculate that few of the visitors realize they are climbing the inside wall of a volcano, and that the bay behind them is the flooded crater left by what may have been human history's greatest eruption, a blowup that could have helped destroy Europe's earliest civilization.

With charm, grace, and peaceful power, this Minoan civilization, as we call it, ruled the island-dotted Aegean Sea 37 centuries ago. Its people arrived on Crete about 3000 B.C. By around 1600 B.C., when Babylon was a leading power, they had learned to write, but to this day their script, called Linear A, has defied all efforts to decipher it. As a result, we know nothing certain about the Minoans' origins.

Santorini: Life on the Rim of Cataclysm

Flowers guard the lip of a bay where a mountain once towered 900 m (2,950 ft) above sea level. Ancient Thera, as Santorini is also known, exploded so violently that its cone collapsed, leaving only a shattered, ring-shaped remnant, visible here in the foreground and as encircling headlands 10 km (6 mi) away. Thrusting from the caldera's center, an island of new lava attests to the volcano's continuing activity. A 1956 earthquake demolished 2,000 houses on Santorini in less than a minute. The blue waters of the Aegean fill the caldera to depths of nearly 400 m (1,300 ft).

OVERLEAF: *The town of Phira, Santorini's largest settlement, perches atop 240-m cliffs sheared by the ancient volcano's massive eruption. Alternating dark and light bands of rock represent lava flows and ashfalls—the internal layers of the ancient cone. Ash quarried from the island goes into concrete mixes and building materials; the Suez Canal is lined with Santorini cement.*

2ND OVERLEAF: *Snow-white walls, terraces, and zigzag stairways in the village of Oía interlock and buttress one another—creating a picturesque architectural style suitable for steep Santorini slopes.*

The comforts of home grace a Santorini cave dwelling. Thera's eruption left a blanket of ash and pumice up to 60 m (200 ft) thick on remaining fragments of the island, material so workable that some islanders dig homes into hillsides.

But thanks to decades of archaeological work, mostly on Crete, we do have examples of their elegant pottery, beautiful jewelry, and artisans' tools. Their palaces and villas, lavishly decorated and outfitted with bathrooms and indoor plumbing, leave no doubt that this was a brilliant and sophisticated society.

It was also a peaceful one. Battle scenes hardly occur in Minoan art, and none of their palaces or busy harbor towns were walled. The Minoans traveled far from their homeland in Crete, trading in crafts and crops, copper and tin. Their vessels crisscrossed the Aegean and voyaged as far as Egypt and Italy.

That was the situation in 1500 B.C. Yet only 50 years or so later the Minoan towns on eastern Crete lay destroyed, their populations dispersed. The prosperous trading empire had been abandoned, and the palace at Knossos, seat of the greatest Minoan ruler, was taken over by Mycenaeans from the Greek mainland. Minoan civilization had been dealt a mortal blow.

But by what means? British archaeologist Arthur Evans, the man who uncovered the Minoan palace at Knossos in the early 1900s, blamed the catastrophe on a series of earthquakes followed by fires—a likely enough event in the tectonically active Aegean, where the African Plate dives beneath the Aegean seafloor. But the explanation didn't satisfy Evans's Greek colleague Spyridon Marinatos.

In 1939 Marinatos suggested that a volcanic eruption, along with earthquakes, tsunamis, and a crop-killing ashfall, desolated Crete. The volcano, he said, was Santorini, only 113 kilometers (70 mi) north.

Ashfall and pottery leave an archaeological puzzle

Santorini, or Thera, once a single volcanic island, blew out its center in the 15th century B.C., then collapsed into the sea.

Krakatoa's explosion blew out 18 cubic kilometers of debris and shook houses 750 kilometers away —farther than the distance from Boston to Washington, D. C. Thera's final explosion, by comparison, blasted out perhaps five times more material and left a caldera four times as big. Its ash layer—after 3,500 years of erosion—is still 60 meters thick in places.

For lack of evidence, Marinatos's linkage of the Theran eruption and the desolation of Crete went largely ignored for decades. Then deep-sea core samples from the Mediterranean east of Crete revealed a layer of volcanic ash from Thera, evidence that the ash must have fallen on Crete as well. And in 1967 Marinatos himself dug into Santorini's ash and pumice at Akrotiri; almost immediately he found part of an entombed Minoan city.

On a visit to the Akrotiri excavation, now roofed against the weather, I saw evidence that an earthquake had wrecked a staircase in one of the ancient stone houses. Later, Minoan work crews had arrived to begin repairs, but a rain of pumice forced them to flee and leave some of their tools behind. The ashfall buried the ruins, preserving pottery and several exquisite wall frescoes.

With the Theran eruption and the destruction of Minoan Crete so close together in space and time, some scientists and historians are tempted to call the two events

cause and effect. But the evidence is only circumstantial, and there are equally convincing arguments that Minoan civilization survived the eruption. Very little ash has been found in archaeological digs on Crete—not enough to have caused major crop failures. Nor is there evidence of tsunami damage on many of the Aegean islands.

Then, too, no pottery unquestionably of the latest Minoan "marine" style has so far turned up at Akrotiri. A lot of it was made on Crete in the 50 years just before the end. Surely, say the doubters, some of it must have found its way to Akrotiri—unless Akrotiri was already buried while Crete still lived. The volcano, they argue, must have erupted half a century before the Minoan collapse on Crete.

But the pottery evidence is still too inconclusive to support either camp, and no one has yet set a convincingly firm date for the eruption. So the debate among archaeologists, volcanologists, and others comes down to questions of likelihood and timing so fine that existing technology has not resolved them.

Further digging may help solve the mystery of the volcano's impact on Minoan history. But since Marinatos's death in 1974, excavation at Akrotiri has slowed, and if Santorini's ash hides other secrets, they may be a long time coming to light.

Exodus, plagues, and lost Atlantis

Furthermore, scientists have even wider effects of the Thera eruption to investigate. For example, most scholars have dated the Exodus of the Children of Israel at around 1200 B.C. But a new translation of an Egyptian inscription that may

Ruins of the labyrinthine palace of Knossos, possible seat of Minoan power on the island of Crete, hint of past grandeur. Minoan society disappeared suddenly, leaving a legacy of archaeological treasure such as this frieze (right) from a sarcophagus, showing priests and priestesses offering gifts and libations to the gods.

Perched above a subduction zone, Crete suffers devastating earthquakes; several damaged the Minoan palaces. Minoan civilization may have toppled from a one-two punch, by both tremor and volcano, for the subducting African Plate fuels an arc of fire across the Aegean: volcanic isles like Mílos, Kos—and Santorini.

316 refer to the Exodus suggests that the departure from Egypt took place nearly three centuries earlier, about 1477 B.C., during the reign of Queen Hatshepsut. If the new date is correct, it may have been a Theran tsunami that drowned the pursuing Egyptians as the Israelites fled along the Mediterranean coast at the beginning of their 40-year trek to the Promised Land.

And volcanic effects well documented in modern times could have caused some of the ten biblical plagues that struck Egypt during the time of Moses. Dust and ash could have darkened the heavens for days on end, as happened with Krakatoa; toxic substances raining down could have poisoned cattle and man, as happened with Lakagígar, or might have stained the rivers blood red with algae and provoked desert-dwelling toads and insects into a frenzy of activity.

The Thera eruption may also lie at the heart of one of the greatest legends of all—that of lost Atlantis. Passed on to us by the Greek philosopher Plato, perhaps from older sources, it tells of a rich island power, "a great and wonderful empire" that was swallowed by the sea in a day and a night, leaving only shoals of mud. Some scholars now point out that a translation error increased the size of the Atlantis Plato described by a factor of ten; corrected, the proportions suggest an island or islands that could fit in the eastern Mediterranean.

There are other parallels, too: According to the legend, the Atlanteans possessed a precious substance (Minoans smelted bronze from copper and tin). There were hot springs on Atlantis (as there probably were on volcanic Thera). And Atlantis was the pathway to other islands and, beyond them, to a continent (just as Santorini lies between Egypt and Europe).

Other details of Plato's description of Atlantis are harder to match. But it is not impossible that the legend of Atlantis reflects an old memory of real events, for many Minoans survived the collapse of their civilization. Some settled along the coasts of the Mediterranean, bequeathing a rich cultural heritage throughout the Greek and Middle Eastern worlds. Minoan pottery and jewelry have been found in Mycenaean tombs in Greece, and the Minoan religion helped shape the Greek pantheon. Groups of Minoans may have settled in North Africa and Asia Minor, and the mysterious Philistines of the Bible—"the remnant of the country of Caphtor"—probably were Minoans.

Most of the civilization-shaking eruptions we know seemed to affect only one civilization at a time: Merapi and Mataram, Ilopango and the Maya. But if even part of Marinatos's theory proves to be true, then Thera may have swayed the course of history and culture for the Western world. Without the eruption, would the Minoans have continued to thrive? Would the Mycenaeans—forerunners of the classical Greeks—have stayed in the shadows, and would our public buildings display Minoan pillars instead of classical Greek columns? Somewhere, perhaps in the unexcavated ash of Akrotiri or in the ooze of the Mediterranean seafloor, archaeologists and geologists may find answers to at least some of the questions bestowed by antiquity.

A solitary hiker finds quietude where, 100 million years ago, colliding continental crust thrust a crumpled seabed into the sky. Slivers of land from across the Pacific struck North America's coast, raising the rock of these eroded peaks— the Ramparts, in Canada's Jasper National Park.

Fingers of mist comb the hollows of the Great Smoky Mountains, a wooded preserve straddling the border between North Carolina and Tennessee. The Smokies are part of the 2,400-km Appalachian chain that extends in long ridges and valleys from Alabama to Quebec. The range, now worn by erosion, may have stood as tall as today's Himalayas after a shoulder of Gondwana nudged Laurussia 300 million years ago—a span of time that makes these rounded knobs and ridges some of the oldest mountains in the world.

If by some fiat I had to restrict all this writing to one sentence, this is the one I would choose: The summit of Mt. Everest is marine limestone.
John McPhee   **BASIN AND RANGE**

Every winter, howling winds sweep snow and sleet across mountain slopes of North America. Families nestle by their fireplaces, sipping hot cocoa and dreaming of palm trees and sunny beaches a plane ride away.

Little do they realize that land from the tropics has come to them. A cabin-bound dreamer in British Columbia may be sitting on ground that started its northward journey some 200 million years ago as an equatorial island. Lovelock, Nevada, probably had an earlier life as a volcanic island far off the Pacific coast of North America.

In fact, over 70 percent of the land that makes up the mountain ranges of western North America came from somewhere else within those 200 million years. When the dinosaurs walked the Earth, the Rocky Mountains we know didn't exist. What is now desert or forest was then shoreline or continental shelf; at one time the western coast lay approximately along the present-day Continental Divide. Muddy rivers wound their way across the face of the land, disgorging their loads onto the continental shelf, broadening it in the same manner that today's eastern continental shelf grows seaward.

As Pangea split apart, North America drifted westward, widening the newborn Atlantic Ocean as it overrode the vast Pacific Ocean floor. But the Pacific Basin was not a featureless plain that plunged

Earth's skin bends and breaks under the force of collision. When a slab cracks and slides over itself, the result is an overthrust. Here great pressure can push oil and gas up through porous sandstone (dark gray layer) until both are trapped under a vaulted ceiling of impermeable rock (light gray). Along North America's overthrust zones (striped bands on map), collision has moved mountains. Parts of the Rockies in Montana have been shoved 65 km eastward, parts of the Appalachians 100 km westward. Computers now aid in the search for deep pockets of oil and gas in both overthrust belts.

smoothly beneath the advancing continent, gone without a trace. It carried island and seamount chains, shallow oceanic plateaus, and microcontinents, just as it now holds the Hawaiian Islands, the barely submerged Fiji Plateau, New Zealand—perhaps 25,000 islands in all.

Some geologists speculate that many of those earlier features were remnants of a lost continent they call Pacifica. They suggest that Pacifica broke off from Gondwana about 220 million years ago and drifted into the Pacific Ocean Basin. There it broke into at least five major chunks and probably many smaller pieces.

Other geologists doubt the existence of Pacifica, but agree that much of the western half of North America once belonged to some other continent—or to no continent at all. They say, for example, that about 100 million years ago an island-arc array called Wrangellia drifted northward and collided with North America, some blocks sliding onto the continent, others continuing north along a strike-slip fault similar to the San Andreas. By about 40 million years ago Wrangellia's pieces, as well as many smaller bits of crust, had cemented themselves onto the continent from eastern Oregon to central Alaska.

North America felt at least 50 such collisions as the subducting ocean floor deposited its passenger islands and continental baggage. These blocks twisted and turned as they merged with the continent.

Compressed by the force of the collisions, rock in the old continental shelf folded and faulted, piling up on top of itself. Chunks broke off and rammed eastward, riding up

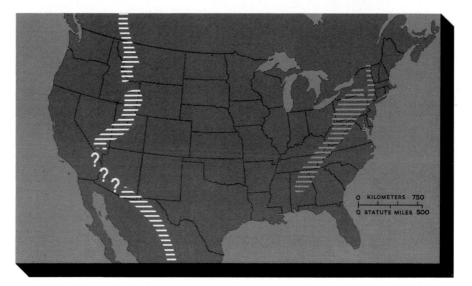

322 and over younger sedimentary layers. The normal rule of stratigraphy—deeper means older—was reversed. Old rock sat on top of young: an overthrust belt. Locked in these layers were the remains of a profusion of marine life.

Within the compacted sediments, pressure and rising heat slowly cooked the organic remains into a brew of gas, salty water, and tiny oil globules. Under pressure, oil and gas oozed into permeable rock, then migrated upward to zones of lower pressure. But often an overlying roof of impenetrable shale or other rock would halt this motion, damming the oil and gas in traps of porous rock, most often sandstone.

For ages these treasure troves remained locked in deep subterranean chambers from Canada to Mexico, beyond the reach or ken of humans. The search went on thousands of feet above. Occasionally it was successful, as when prospectors struck natural gas in Alberta in 1913. But in the United States nearly every test hole in these overthrust areas came up dry. Finally, a decade ago, the luck of the exploration geologists changed.

Not their luck, really. Their equipment. The microelectronics revolution of the 1970s gave geophysicists computer power to help them interpret seismic information about overthrust belts far more accurately than they could before. Using this information, they pieced together pictures of the rock layers, searching as deep as 6,000 meters (20,000 ft) for oil and gas traps. In 1975, one company finally hit a pay zone in northeastern Utah: 540 barrels of oil and 270,000 cubic feet of gas rose out of the hole each day. A new rush for black gold was on—and the search itself provided even more details about the stratigraphy of the western United States.

Oil geologists now estimate that the quantity of oil beneath the western overthrust belt may equal that of Prudhoe Bay in Alaska. The potential in Wyoming, Idaho, and Utah alone could equal more than a year's supply of oil and nearly four years' supply of natural gas for the entire United States.

Overthrusting is not unique to the Rockies. An older collision, between converging continents, also produced the phenomenon. Discovery of oil in the Rockies led petroleum companies to pursue their search in another overthrust belt. They turned east, to the Appalachian Mountains.

More than 300 million years ago, when Pangea was forming, Europe and Africa began to bump into North America. As they did, a block of crust called Avalon was working its way up the eastern shore the same way Wrangellia later slid north along the western coast. When Africa finally closed the Iapetus Ocean, forerunner of the Atlantic, the Appalachian chain had begun to rise. By Permian times these mountains stood high and mighty, probably looking much as the Himalayas look today. The Avalon chunk had fixed itself securely to the North American continent. Pieces of it appear today from the Carolinas to Newfoundland.

The collision formed an overthrust belt that runs for 1,800 kilometers (1,100 mi) along eastern North America. Trapped organic sediments from the Iapetus Ocean are widespread beneath layers of

OVERLEAF: *The Garden Wall, a blade of sedimentary rock honed by the millennia, slices across Montana's Glacier National Park, a geologic layer cake laid down on a seabed and shoved here in an overthrust. Glaciers gouged this chunk into a precipitous ridge; weather may someday finish the job and wear it away entirely.*

A maelstrom of colliding plates and continents roils the Mediterranean region. As the African Plate converged on the Eurasian, it jammed Italy into Europe, creating the Alps. Spain pivoted southward, pinching up the Pyrenees as it tore away from the Atlantic coast of France. Pushed westward by Arabia, Turkey gradually invades the Aegean.

The crush of collision can wrinkle layers of sedimentary rock into S-shaped recumbent folds. Sichelkamm, a peak in the Swiss Alps (lower), keeps only the bottom half of the S; erosion has sheared off the top.

326

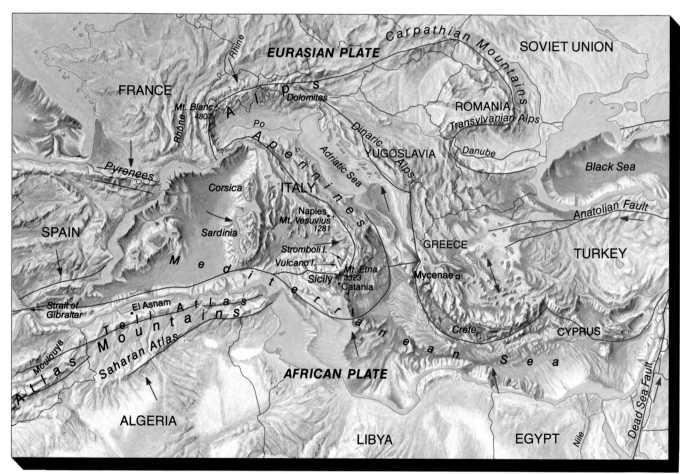

□ Archaeological site

rock extending from Pennsylvania to Alabama—the larger part of the eastern overthrust. Indeed, the oil industry in the United States began in western Pennsylvania when Edwin L. Drake found a shallow pocket of oil in 1859. Now, armed with new knowledge from plate tectonics, oil companies drill much deeper in hopes of renewing their strikes.

Folding into the sky

Europe too shows the effects of continental collision. The upper layers of the crust have been folded and twisted to form the Alps, Apennines, and Carpathian mountain ranges.

How can hard, solid stone bend? The layers of sedimentary rock in the Alps originally formed on a seabed. They have never melted, yet now you can see them folded into tortured loops in the cliff faces of peak after peak.

Paul Tapponnier of the University of Paris specializes in the study of rock deformation. I met Paul during a series of dives we made in the 1970s. He explains that rock is plastic: It will bend with heat and pressure, or with pressure and time. All three are at work deep inside a mountain range.

Under the "pressure" of gravity, even hard, cold glass will deform with time, Paul says. "If you look at the stained-glass windows of Notre Dame, which were set in the 13th century, you'll see each pane is now thicker at the bottom than at the top. The glass has flowed downward over the centuries.

"So time is like heat. If pressure is applied for a very long time, stone can deform without melting."

If we could duplicate the heat and pressure under a mountain range being squeezed between colliding plates, we could, given enough time, fold a tombstone. This is what has happened to once level strata in the Alps and in collision zones throughout the world.

We can think of tectonics in the Mediterranean Basin as a smaller version of plate activity worldwide. Several small plates pinched between the African and Eurasian Plates crash, grind, separate, and rotate into one another. The result is geologic action on a grand scale—and human disaster.

Continuing collision and subduction of these plates, with accompanying volcanism, buried Pompeii, shattered the peace of Minoan civilization by blowing up the island of Santorini, and inflicts modern calamities—earthquakes in Greece, Algeria, Italy, Romania, Yugoslavia, Morocco. . . .

Today's Mediterranean geology stems from the period shortly after the breakup of Pangea, when a larger body of water, part of the Tethys Ocean, occupied the area.

As Africa and Eurasia slowly converged, the floor of the Tethys Ocean vanished beneath the more buoyant continental blocks. But sediments on the continental shelf were thrust upward from beneath the sea. The most spectacular result of this head-on encounter is the Alps, a range sweeping out in an arc 1,000 kilometers (620 mi) long and 160 kilometers (100 mi) wide.

Pioneering scientists scaled the jagged Alpine peaks to make studies of their structure and to collect rock samples. From these hard-won observations, a complex picture slowly began to take shape.

Swiss geologist Rudolf Trümpy has identified five short stages or pulses of mountain building during the past 100 million years. Between pulses, erosion did its work; as recently as 5 million years ago there were no mountains left here at all. The present-day Alps began to rise about 3 million years ago—and most geologists agree with Trümpy that they are still rising, perhaps one or two millimeters a year. (Unlike many collision zones, however, the Alps have no active volcanism.)

The Alps barred the heart of the continent from the gentler climate to the south. They gave birth to the Rhône, Rhine, Adige, and Po, major rivers all. And they helped shape European history; the armies of Hannibal, Julius Caesar, and Napoleon pursued victory through the gateway passes of the Alps.

Today we regard this range as one of the most beautiful places on Earth. But to many people in the 18th century, the mountains were ugly, even accursed. Superstitious peasants whispered of dragons roaming the valleys. Travel there was difficult—and so thrilling that it produced in one visitor "an agreeable kind of horror."

Even as the Alps grow, erosion keeps growth in check. Streams etch deep valleys into their flanks, and on the heights more than 1,200 glaciers patiently carve the rock.

Spectacular peaks such as the Matterhorn and the Eiger have long challenged climbers. Mont Blanc, highest of the Alps, defied mountaineers until 1786, when a couple of French climbers finally reached its 4,807-meter (15,771-ft) summit.

With macabre serendipity, the

death of some later climbers taught scientists something about the rate of movement of Alpine glaciers. In 1820 an avalanche of ice and snow buried a team of mountaineers in Glacier des Bossons. In 1861 their frozen corpses emerged from the snout of the glacier 3,165 meters (10,384 ft) below; they had advanced at a rate of about 22 centimeters (8.5 in) a day.

Seafloors on high

The island of Cyprus stands like a gem above the eastern Mediterranean. Its name reflects its worth, for the word "copper" comes from Cyprium, the Roman name for the island, which came in turn from the Greek name Kypros. For thousands of years, the principal export from Cyprus was copper. Extensive pre-Roman mining works remain today at Skouriotissa.

Cyprus has copper because much of the island is ancient seafloor—and seafloors can hold rich deposits of metals such as copper, a legacy of seabed formation at the Mid-Ocean Ridge. In Cyprus the old ocean floor has been piled up in a geologic formation called an ophiolite. The word comes from the Greek *ophis,* serpent, because the rocks in an ophiolite are a blotchy green, like many snakes.

As Africa and Eurasia caught the ocean floor between them, some of it was scraped off the subducting plate. All over the world, collision zones both active and ancient hold ophiolite slivers.

Geologists hunt for ophiolites not only for mineral wealth, but for the secrets they hold about Earth's behavior. Ophiolites are samples of the ocean floor's internal structure.

Some of them are older than any present-day seafloors; they preserve samples of ocean lithosphere that has long since been lost to subduction. A few hold a complete sequence of seafloor building: dark, heavy rock that once was the bottom layer of the lithosphere; above that the hardened magma that filled the chambers of the Mid-Ocean Ridge; next the magma that solidified enroute from chamber to seafloor in cracks opened by plate motion; then the pillow lavas that formed on the seabed; and finally the sedimentary rocks laid down in the ancient sea.

Studying the ocean floor through the windows of tiny submersibles, as I have, can be an expensive and often frustrating experience. Crude mechanical arms make poor substitutes for hands. But on an ophiolite we can do without our submarines and still inspect the seafloor—even examine it in cross section.

The world turned inside out

In early 1981, one of the world's leading experts on ophiolites took me to a stretch of exposed ocean floor in the Sultanate of Oman, on the toe of the Arabian Peninsula. We drove across the dusty desert, through villages of low, white, shuttered houses, past a bustling marketplace, and on to the rocks.

My guide was Clifford A. Hopson, an old friend and diving colleague from the University of California at Santa Barbara. When I had attended UCSB as an undergraduate, Cliff was my professor in geochemistry. Now, 17 years later, he was still teaching me geology.

The ophiolite of Oman, like that of Cyprus, was caught when the

From Seafloor Deep to Shining Peak

Switzerland's Lauterbrunnen Valley makes an idyll out of a battleground of colliding continents and grinding ice. Here in the Alps, rivers have carved V-shaped valleys into the landscape; glaciers have bulldozed some into a broad U shape. Today visitors to the village of Lauterbrunnen hike to the base of Staubbach Falls and gaze up through its mist to a cliff top that may have been the valley floor before the glaciers came.

OVERLEAF: *Muffled year round in wintry coats, the peaks of Switzerland's Bernese Alps began as sediments under the warm Tethys Ocean, ancestor of the Mediterranean. Italy's push into Europe focuses on this central Alpine region, raising the taller summits more than 4,000 m above sea level. And the thrust continues; each year these mountains grow a little higher.*

2ND OVERLEAF: *Ramparts of the Dolomite Mountains stand watch over golden meadows. These mountains take their name from dolomite, a hard, acid-resistant limestone formed on seafloors. The mineral, in turn, takes its name from Déodat de Dolomieu, an 18th-century French mineralogist who first described its properties. Fossilized sea creatures found in the region are sold in local souvenir shops.*

Bulbous blocks of pillow lava, formed under the sea some 95 million years ago, lie exposed on the desert slopes of the Hajar Mountains in Oman. The collision of the Arabian and Eurasian plates pushed this piece of seafloor to its new high and dry location.

A marine reptile that swam the shallows of the Tethys Ocean more than 200 million years ago came to light on a Swiss mountainside. This fossil skeleton of a foot-long Pachypleurosaurus, *one of several found on Monte San Giorgio, was discovered in a layer of sedimentary rock rich in organic material from the ancient seabed.*

Tethys Ocean gave way to converging continental crust, in this case Arabia and Eurasia. The seafloor between them cracked at its weakest point—the Mid-Ocean Ridge where the two plates met. Ocean floor rode up onto the Arabian Plate and carried up with it a segment of the ridge itself, including the magma chamber. Now both the ridge and the chamber were stranded on dry land. Erosion had cut into this ophiolite, leaving a cross section of the floor of the Tethys Ocean.

Late in the day, as the sun sank below the jagged hills, we stopped the Land Cruiser. A few spiny acacia trees stood here and there. I gazed around at the same rocky beige desert I had seen all day. But it was a moving sight for a geologist, for at that moment we were right in the middle of the old magma chamber, once the inside of the Earth.

The final stop on this journey to the seafloor was at the Oman Mining Company property near Sohar, reputed birthplace of the legendary Sindbad the Sailor. Here, with the help of the Saudi Arabian government, the Sultan of Oman is trying to diversify Oman's oil-dependent economy by opening a copper mine. The entrance to the mine begins on the ancient Tethys floor. In the surrounding desert are pillow lavas like those I have seen countless times through the viewports of *Alvin* along the Mid-Ocean Ridge. Nearby stand the remains of an ancient smelter, a reminder that mining has gone on here for more than 5,000 years. Some people say that puts Oman among the contenders for discovering the art of smelting.

Descending the mine shaft, we walked deeper and deeper down the dark ramp, the lights on our mining helmets illuminating the glassy lava surface overhead. Some 175 meters down we had to halt; groundwater leaking in from above had flooded the shaft. Once the water is pumped out and the shaft excavated to its working depth of 225 meters, mining the copper will begin.

The zone of collision between Africa and Eurasia has provided metal to build civilizations, but it has also created agents of destruction.

Fiery lava causes armed battles

"For centuries the population living on the slopes of Etna have accepted the violence of the volcano as something imposed by that relentless fate from whose control there is no escape. Only divine intervention can change the course of fate and if man has deserved the intervention of this Patron Saint, the fury of the volcano will stop."

So wrote Dr. Lillo Villari, Director of the Istituto Internazionale di Vulcanologia, located at the base of Mount Etna in the Sicilian city of Catania. When I visited him in 1982, he spoke of the constant danger Mount Etna poses to the people crowding its slopes.

Etna and other Italian volcanoes owe their vigor to the collision of the African and Eurasian Plates. A segment of the African Plate is descending beneath the tip of Italy's boot, creating a subduction zone.

Stromboli and Vulcano, two active volcanoes on the Lipari Islands north of Sicily, tap the magma generated by this subduction. Stromboli has been called the "lighthouse of the Mediterranean" for the glow from its summit. It has been almost continually active for more than

339

2,000 years and shows no sign of turning off its light. But volcanologists fret even more about Vulcano, for every summer thousands of tourists pour onto the island to bask in a volcanic mud bath at its base.

These volcanoes, products of the subduction zone below, have simple origins compared to that of Mount Etna, where the process of collision has reached an impasse. Pulled and pushed in a number of directions at once, Etna has developed complex fracture patterns. Often it erupts along a fracture on its flank instead of from the summit. This can give residents little time for building barricades to divert the flows. It may be just as well, for diversion attempts themselves have proved dangerous.

On Mount Etna's densely populated base, deflecting a flow of lava away from one community frequently means directing it toward another. "The inhabitants of the neighboring villages, feeling potentially threatened by any diversion of the lava flows, have not hesitated to take up their arms," says Villari.

Dr. Gianni Frazzetta of the institute took me to see the snowcapped 3,323-meter (10,902-ft) summit of Etna on a beautiful spring day. We drove out of Catania, a town that Etna has already devastated seven times (the latest in 1669). We passed through suburbs terraced with volcanic rock and emerged onto a bare lava slope. There were no plants, no signs of animal life. We came to the snow-covered middle slopes and a ski lodge. From there we continued by ski lift.

Rolling clouds cloaked Etna's simmering summit. In September 1979, nine tourists climbing near here were killed by a sudden eruption that bombarded them with glowing rocks and ash. That part of the path is now closed to visitors.

It was balmy down in Catania, but up here the wind was so bitter we gave up trying to reach the summit. As we turned back, I stopped for a final picture, only to find my camera shutter frozen.

To find Italians living on volcanic slopes is nothing new. Two thousand years ago the Greek geographer Strabo wrote, "Mount Vesuvius . . . save for its summit, has dwellings all around, on farmlands that are absolutely beautiful."

I rode a cable car to the summit of Vesuvius with a group of Parisian fashion models. When we reached the top, their high heels made it difficult for them to walk about the crater and down the zigzag path to the smoking fumaroles. In 73 B.C., the gladiator Spartacus had used this crater as a fortress for his band of rebel slaves and gladiators. Now we looked down on the excavated ruins of Pompeii, destroyed by an eruption in A.D. 79. Pompeii, like other settlements on the slopes of active volcanoes, grew here because of the rich volcanic soil. According to Strabo, that soil yielded grapes, olives, and up to four seed crops a year.

Pompeii and nearby Herculaneum were sealed as if in a time capsule under the eruption's blanket of ash. Archaeology has made them world-famous, yet they were small compared to the population that today lives within reach of Vesuvius. And Vesuvius is only one of the many volcanoes in the Naples area, where millions of people live.

Amid the massive Neapolitan

A survivor of a 1980 quake that demolished El Asnam, Algeria, salvages belongings from the wreckage of his home. The tremor struck without warning on October 10, leveling most of the city and claiming some 3,000 lives. "The dogs didn't have time to bark," recalled one survivor. "It was all over within seconds." Sited near faults caused by the converging African and Eurasian Plates, El Asnam is no stranger to severe jolts: A 1954 shock ravaged the city almost as thoroughly as this one.

rush hour, I left the city one morning for the town of Pozzuoli, just to the west. On the way I crossed Agnano crater, some two kilometers across. People and buildings pack it from wall to wall. In the crater is a hospital and health spa. Hot-spring water flowing from the walls of the crater collects in a series of vats where its hot, mineral-rich mud can settle out. Here I found workers collecting the mud; the nearby hospital uses it to treat arthritis, rheumatism, skin ulcers, and respiratory ills. Some 2,000 years ago the Romans built steam rooms over many of these hot-spring sites. Today workers renovate them for modern Italians to enjoy.

Agnano crater was once famous for its "dog grottoes." The Romans entertained tourists by throwing dogs into these underground chambers, where the animals would suffocate in an invisible layer of toxic volcanic gas. Fortunately for the dogs, the grottoes have been covered over.

I continued to Pozzuoli, a town in a large, active volcanic region of many lakes and craters known as Campi Flegrei, or "fiery fields". Since Roman times, Pozzuoli has been known to slowly rise and fall. A temple ruin now lies partly submerged off the city's waterfront, its columns acting like depth gauges as they rise and sink. Some scientists believe this motion comes from the inflation and deflation of a large magma chamber beneath the city. In 1970 the ground had risen three feet in less than six months; many buildings were evacuated as unsafe.

Volcanoes may threaten the people of southern Italy, but earthquakes have claimed more lives in recent years. On November 23, 1980, a quake measuring between 6.8 and 7.0 on the Richter scale destroyed many towns inland from Naples. More than 4,000 died.

Most of the demolished buildings had been built of stone more than a century ago. But not all; in Sant' Angelo dei Lombardi a three-year-old hospital, built to the latest earthquake standards, collapsed and killed scores of people. The saddest story I heard was the loss of a hundred of the town's young people. They died at a dance when the building collapsed on them.

I was not surprised to find that the lack of progress in restoring the cities had become a national issue. In Sant' Angelo, half the people displaced by the quake still lived in trailers clustered along the rubble-strewn streets. The town looked like a war zone.

The greatest collision shatters a continent

To many of us, plate tectonics at first seemed a simple matter. A few large rigid plates floated majestically about on the surface of the Earth, spawning quakes, volcanoes, and mountain ranges where their edges met. But as earth scientists work out the details of plate tectonics, we have discovered that the effects when plates meet are not that simple; complex events can happen far from their edges as well.

Nowhere is this complexity more spectacular than where India hits Asia: the Himalayas. Behind the Himalayas stretches the Tibetan Plateau, where the *average* elevation is higher than that of any mountain in the 48 contiguous United States. And beyond Tibet sprawls China, scene of earthquake-prone faults, climatic extremes, and a unique geography. Paul Tapponnier and his frequent collaborator, Peter Molnar of the Massachusetts Institute of Technology, believe the forces that have developed here, as one continent tries to push another out of its way, may be spawning earthquakes and shaping landforms thousands of kilometers from the point of collision.

Since 1177 B.C., Chinese accounts have described more than 3,000 destructive earthquakes. The old regimes kept detailed record books, since earth tremors were thought to reflect a ruler's standing with heaven. Modern seismologists mine these records for their accurate seismic history.

For example, in A.D. 1303, a huge quake, since estimated at 8.0 on the Richter scale, hit Shanxi Province in east-central China. The accounts say that "the ground opened up as if it were a canal. Water was spouting and sands boiling. One thousand four hundred palaces and temples collapsed utterly." The deadliest earthquake recorded in human history struck neighboring (and similarly named) Shaanxi Province in 1556, killing 830,000 people.

Tapponnier and Molnar believe many of China's earthquakes relate to the forces that push up the lofty Himalayas far to the southwest.

Home of the snow
The Himalayas, a crescent-shaped barrier 2,500 kilometers (1,550 mi) long, rise at the northern edge of the Indian subcontinent. Thirty of the world's highest peaks stand here,

Baykal Rift

Tangshan

In a collision that began at least 40 million years ago, India has rammed 2,000 km into Eurasia and piled up the Himalayas and the lofty Tibetan Plateau. Pinned against Siberia, parts of China and central Asia are squeezed aside. Arrows on edge show motion along major faults (lavender strips) like the Altyn Tagh; those lying flat

show motion of crustal blocks. As the blocks carom and jostle, one pair far from the impact zone are knocked apart, opening up a rift (plunging arrow) that cradles Siberia's mile-deep Lake Baykal, the world's deepest lake.

including the 8,848-meter (29,028-ft) Mount Everest. Among the people of Tibet, Mount Everest is revered as Jomo Langmo, "goddess mother of the world." The name Himalayas derives from Sanskrit, *hima* meaning "snow" and *alaya* meaning "abode."

Unlike the Appalachians, the Himalayas are still very much alive. Some 70 million years ago, after the breakup of Gondwana had split India away from Africa's eastern flank, India began to drift northward. The drift became a tectonic sprint toward the southern rim of Eurasia at about 10 centimeters (4 in) per year. Some 40 to 60 million years ago, the two continents met. The force of the collision slowed India by half and pushed a band of oceanic crust from the subducting plate up into the suture joining the continents. As India continued to plow into Eurasia, compression piled up the Himalayas—including that seafloor.

These mountains contain a complex mixture of rock welded together by the collision. Sedimentary strata contain fossil shells of tiny marine animals that once lived in the warm Tethys Ocean between India and Asia. Their remains were caught in the squeeze and incorporated into the rising range. Born in a tropical sea, they now lie under a perpetual layer of ice and snow at the ridgepole of the world.

The Himalayas control the region's weather by keeping most of the monsoon rains from reaching Tibet. Laden with moisture from the Indian Ocean, the monsoons sweep northeastward over India and are forced upward by the lower slopes of the mountains. The mois-

ture condenses and falls in downpours that erode the lower elevations. Snows on the heights feed glaciers that carve the mountains into sharp, rugged peaks.

Yet, despite this onslaught, the Himalayas remain the world's highest mountains because they are still rising. As the Indus, Ganges, and Brahmaputra Rivers carry away the scourings of the ranges, the ongoing collision forces them upward.

Raising the roof of the world

The subcontinent of India has penetrated 2,000 kilometers (1,200 mi) into the Eurasian Plate. The buckling and thickening of the crust beneath the Himalayas and neighboring ranges has absorbed some of the incursion. But what about the rest?

In the 1970s Tapponnier and Molnar began to examine new images of Asia—satellite photographs that, for the first time, pictured entire regions in a single view. They saw topography and fault systems that hinted of the answer: Not only was the crust thickening, but much of it was being squeezed eastward.

Seismic findings helped fill in the details. Tibet appears to be transmitting the shock of India's impact to China, where blocks of continental crust are breaking up to give India room to move north. Tapponnier and Molnar see Tibet as a pressure gauge, its height suggesting the force of India's push.

Some scientists say Tibet may be the fastest rising major landmass in the world. They estimate that each year the Tibetan Plateau rises half a centimeter (about an inch every five years). This prolonged uplift has changed the Tibetan climate, as monsoons bring less and less rain to

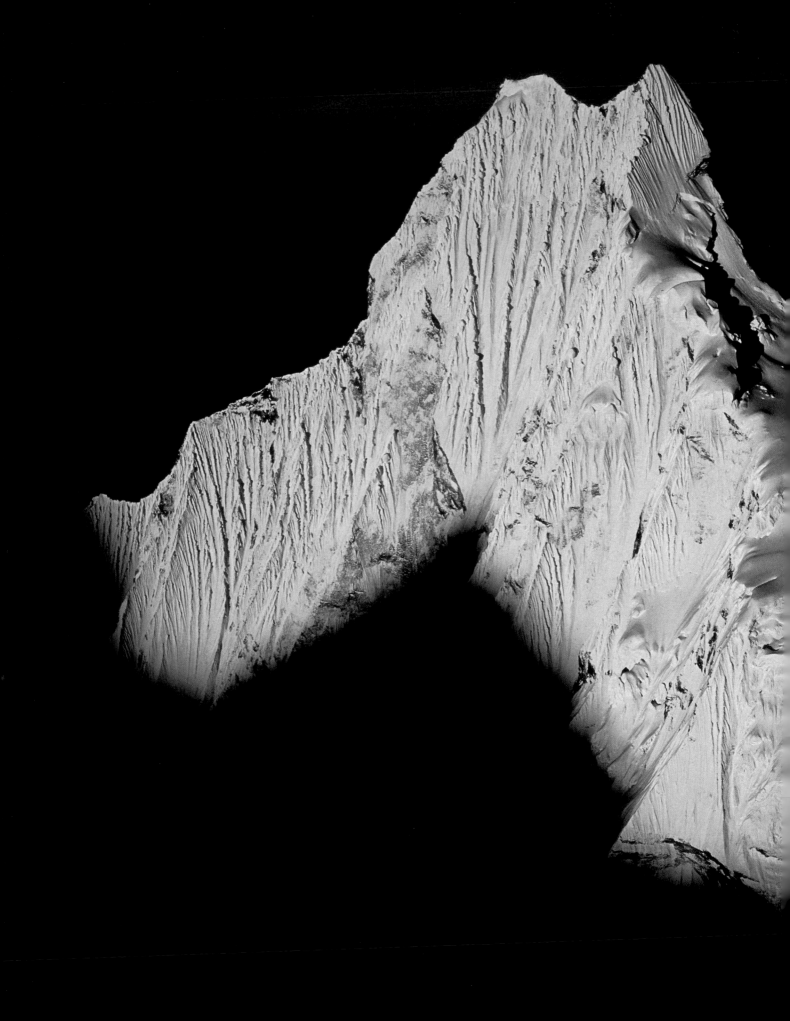

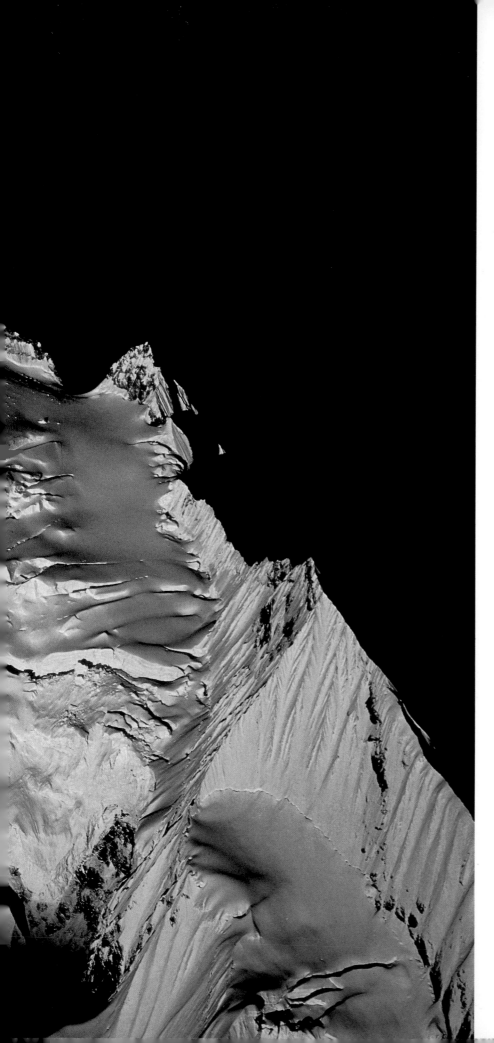

Himalayas: Mountains to Challenge the Soul

Stupendous Machapuchare, part of the Annapurna Range in Nepal, glows with the first light of a new day. The peak, compared by some to the Matterhorn, soars a mile higher than its Swiss look-alike.

OVERLEAF: *"Whirling clouds that blow up from nothing, then just as quickly disappear," wrote climber Reinhold Messner of this scene on a shoulder of Mount Everest after a 1978 expedition. Rising to 8,800 m (29,000 ft), the Himalayas wall off Tibet from the wet monsoons that sweep India. On a wider scale, the sky-piercing bulk of mountains and plateau affects the west-east flow of the jet stream like a submerged rock in a creek bed. Deflected into north-south meanders, this river of air high in the stratosphere influences climate over much of the Northern Hemisphere. Cold snaps and heat waves in North America can sometimes stem from these disturbances half a world away.*

2ND OVERLEAF: *Inching along a fracture "trail" at the 5,500-m level, a climber traverses a face of the Trango Towers in Pakistan's Karakorum Range. As continental impact raises these mountains, glaciers (background) join forces with wind and water to cut them apart. Result: the Towers, jagged granite spires with walls more than a thousand meters high.*

Namche Bazar, Nepal's gateway to Mount Everest, nestles in a glacier-carved amphitheater some 3,440 m (11,290 ft) above sea level. Branches festooned with prayer flags adorn a Buddhist shrine high above the village. Long a marshaling area for climbing expeditions, Namche Bazar is home to the "Tigers of the Snow"—Sherpa porters and climbers renowned for carrying heavy loads at high altitudes. In May 1953 Sherpa Tenzing Norgay and Edmund Hillary of New Zealand reached Mount Everest's summit—the first climbers to scale Earth's highest peak.

the region. Uplift has raised some farming villages so far above the water table on which they depended that people have abandoned them.

Today, the lowest elevation of the Tibetan Plateau is 3,600 meters (12,000 ft). More than 17,000 glaciers cover its surface, making it the "third polar region of the world."

Tibet is so high it tends to flow outward under its own weight, causing small rift valleys to open in the plateau. Paul Tapponnier compares Tibet to soft Camembert cheese, which slowly puddles if left to sit out awhile.

Forced up by a push from the south, the Tibetan Plateau in turn pushes against the landmasses that hem it in. But the stable Siberian Shield and the entire Eurasian landmass keep it from moving much farther north or west. So it moves the only way it can—to the east. Tibet is spilling into southeast Asia, and its push creates a series of strike-slip faults before it. One of these, the Altyn Tagh, is longer than the San Andreas—perhaps longer than any other strike-slip fault on land.

Great blocks of eastern Asia are being squeezed outward toward the Pacific Ocean Basin. This strained tectonic logjam stirs mighty earthquakes in China. The 1556 quake was the most destructive. But one of similar power hit on July 28, 1976, in Tangshan, a city of one million people east of Beijing. From 200,000 to 700,000 people perished, the worst quake toll in modern times.

Yet another of Asia's scars from its bout with India is the Baykal Rift in Siberia. Here, as blocks of crust get wedged to the east, they leave a gap. It has trapped about a fifth of Earth's fresh water: long, narrow, deep Lake Baykal, cupped in a rift opened not by forces from beneath, like the African Rift, but by the side effects of a distant collision.

Next?

And so goes the quest—wresting from the rocks the story of where they have been and where they are going. Today, geologists hunt the world for rifts, faults, spreading zones, and incipient volcanoes, searching for clues to refine the theory of plate tectonics.

We have many mysteries yet to solve: What does drive the plates? Is it convection in the mantle, as many experts now think? What role do hotspots play—and just what *is* a hotspot? How can we use our discoveries to help us solve our energy and environmental problems, and to predict earthquakes and eruptions? Will we ever learn the secrets of our Earth's beginnings?

Whatever it is that moves the plates, we should hope that it continues. For when it stops—when mountains no longer wring out the clouds nor volcanoes rejuvenate the soil—then the mountains will wash into the sea, never to return, and the face of our world will again be "without form, and void." And so we have the central paradox of our living planet—that its most violent and destructive forces make life possible upon its face.

We shall not cease from exploration
And the end of all our exploring
Will be to arrive where we started
And know the place for the first time.
Through the unknown, remembered gate
When the last of earth left to discover
Is that which was the beginning. . . .
T. S. Eliot

353

Glossary

Many scientific terms used in this book are explained in the text and appear in the Index. This Glossary lists some of the most important words, with short definitions for quick reference. Among the sources used was the comprehensive Glossary of Geology, *published by the American Geological Institute.*

356

aa: rough-surfaced, blocky lava.

abyssal plain: deep, flat region of the ocean floor extending outward from the base of a continental slope.

accretion: growth, as of a planet or continent, by external addition of material.

alkaline: having the chemistry of a base (in contrast to an acid). The compounds produced from an antacid tablet are alkaline. Alkaline magma is rich in sodium or potassium, poor in silica.

anticline: an upward fold of rock layers.

ash: loose bits of material under 2 mm across thrown out by a volcano.

asthenosphere: the plastic (semisolid) layer of Earth below the lithosphere; part of the upper mantle.

atmosphere: as a unit of measure, the pressure on any surface equal to that of air at sea level (14.66 lb per sq in). Underwater, pressure increases about one atmosphere for every 10 meters of depth; at 80 meters, the pressure is some 8 atmospheres.

back-arc spreading: crustal spreading behind an island arc; a result of subduction.

basalt: dark igneous rock composed mainly of dense ferromagnesian minerals.

bauxite: a rock rich in aluminum oxides; usually a product of weathering in tropical environments.

black smoker: jetlike hot spring that builds a chimney of metallic deposits on the seafloor.

BP: before present.

calcium carbonate: a compound found in bones and shells; the major component of limestones.

caldera: the large, roughly circular basin of a collapsed volcano.

Catastrophism: generally, the doctrine that sudden violent natural events are the chief agents for modifying Earth's crust.

chert: a hard, dense sedimentary rock composed mainly of quartz; often called flint.

cinders: glassy bits of bubble-filled volcanic ejecta that fall to the ground solidified.

continental drift: an early theory that the continents somehow moved across Earth's surface. Replaced by the theory of plate tectonics, after the discovery of seafloor spreading.

convection: circulation of mantle material caused by heat deep within the Earth.

core: the iron-and-nickel nucleus of the Earth. A liquid outer core surrounds the solid inner core.

craton: stable, ancient area of continental crust.

crust: the outermost shell of the Earth; upper part of the lithosphere.

deformation: the process of folding, faulting, compressing, or extending rock.

diapir: below ground, a body of lightweight material—most often salt—that has squeezed upward under pressure, rupturing overlying layers of rock and rising through them. Underground salt domes, some two kilometers in diameter, are diapirs that have risen as much as ten kilometers.

dip: the vertical angle at which a fault descends into the earth.

dip-slip fault: a fault along which movement is up or down, rather than sideways.

dolomite: a calcium-magnesium carbonate mineral; the type of limestone it forms.

dust (volcanic): fine ash.

ejecta: material discharged from a volcano.

eon: longest interval of geologic time (see chart).

epicenter: the point on Earth's surface directly above the focus of an earthquake.

epoch: interval of geologic time (see chart).

era: interval of geologic time (see chart).

evaporite: any rock formed by evaporation, usually of salt water in a landlocked basin. Examples: gypsum, rock salt, dolomite.

extrusive rock: igneous rock that solidified after reaching the surface.

fault: a fracture in Earth's crust where movement has taken place.

fault block: a piece of crust bounded by faults. Fault blocks form California's Sierra Nevada and Wyoming's Grand Tetons.

fault trace: the line of a fault visible on the Earth's surface.

feldspar: a group of rock-forming minerals that make up 60 percent of Earth's crust.

focus: the initial underground rupture point where rock gives way under stress to produce an earthquake.

fossil: remains or imprint of a plant or animal preserved in Earth's crust.

fumarole: a small volcanic vent from which vapor and hot gases escape.

geyser: a spring that intermittently erupts hot water and steam.

gneiss: coarse-grained metamorphic rock commonly rich in feldspar and quartz.

Gondwana: ancient supercontinent that included South America, Africa, Australia, Antarctica, and India.

graben: a trough created when a block of crust drops between faults.

granite: light-colored igneous rock, mainly of quartz and feldspar; forms when certain types of magma solidify underground.

guyot: a flat-topped seamount (after Arnold H. Guyot, 19th-century Princeton geologist).

horst: a block of uplifted crust bounded by faults.

hotspot: a persistent heat source in the mantle that causes volcanism, often within a plate.

igneous rock: any rock formed from a melt.

intrusive rock: rock, usually igneous, forced into pre-existing rock.

island arc: a chain of volcanic islands over a subduction zone. The arc is bowed toward the subducting plate.

Laurasia: ancient supercontinent that included North America, Greenland, northern Europe, and much of Asia.

lava: molten rock above ground; such rock after it solidifies.

limestone: a sedimentary rock mainly of calcium carbonate; often formed by the skeletons of marine life.

lithosphere: the solid, outermost shell of the Earth; includes the crust and the uppermost portion of the mantle.

magma: molten rock underground.

magma chamber: a reservoir of magma at shallow depths in the lithosphere.

mantle: the thick layer of Earth between the crust and the outer core.

metamorphic rock: any rock changed in composition or texture through heat (without melting), pressure, or chemistry.

mineral: a naturally occurring inorganic material having an orderly internal structure; the basic component of rock.

Neptunism: the theory that Earth's crustal rocks were deposited sequentially by water.

nuée ardente: hot, flowing cloud of ash and gases spewed from an erupting volcano (French: "glowing cloud").

obsidian: dark-colored volcanic glass.

ophiolite: a piece of ocean crust pushed up on land, apparently by plate collision.

orogeny: the process of mountain building.

overthrust belt: an area where part of a block of crust has broken and ridden up over another part.

pahoehoe: smooth-surfaced, ropy lava.

Pangea: a supercontinent of 300 million years ago that included most of Earth's continental landmasses.

period: interval of geologic time (see chart).

planetesimal: a small solid celestial body; according to one theory, Earth formed by accretion of planetesimals.

plate: one of the crustal segments that make up Earth's lithosphere.

plate tectonics: the theory that Earth's lithosphere is made up of separate pieces, or plates, which migrate over the plastic mantle layer.

Plutonism: James Hutton's 18th-century theory that much of Earth's crust solidified from molten rock.

porphyry: any igneous rock composed of large crystals surrounded by small grains.

pull-apart basin: an area where translational movement at a bend in a strike-slip fault zone has stretched the crust, allowing it to subside and form a basin.

pumice: a bubble-filled, lightweight, light-colored volcanic rock; often buoyant enough to float in water.

pyroclastics: anything exploded from a volcano in fragments; ejecta (literally, "transported by fire").

quartz: crystalline silica; after feldspar, the most common rock-forming mineral in continental crust.

rain shadow: an area on the lee side of a mountain range with less rainfall than the windward side. Mountains create a rain shadow by forcing moisture-laden winds high enough for most of the moisture to condense, falling as rain or snow, before the moving air can cross the peaks.

rhyolite: rock formed from viscous, explosive magma; chemical equivalent of granite, but solidified above ground.

rift: a trough or valley formed where two blocks of crust move apart; a large graben.

sandstone: a medium-grained sedimentary rock mostly of quartz.

schist: a foliated rock usually formed from metamorphosed shale.

scoria: pitted volcanic rock, heavier and darker than pumice; large cinders.

seamount: an undersea mountain taller than 1,000 meters.

sedimentary rock: rock formed by accumulation of sediments; usually layered.

shale: fine-grained sedimentary rock, usually formed by compression of clay, mud, or silt.

shield: the exposed part of a craton.

shield volcano: a broad, gently sloping volcano built up by a series of lava flows. Examples: Mauna Loa, Kilimanjaro.

silica: silicon dioxide. Quartz, opal, flint, and most beach sands are made of silica.

silicates: minerals containing silicon and oxygen; they compose most of Earth's crust.

sinter: encrusted mineral deposits left by springs, lakes, or streams; common in geothermal areas.

slate: a fine-grained, foliated rock most often formed from slightly metamorphosed shale.

strata: layers of sedimentary rock.

stratigraphy: the study of rock strata and their geologic history.

strike: the direction of the horizontal axis of a fault.

strike-slip fault: A fault on which movement is to the side, along the strike, rather than up or down.

subduction: the descent of one plate beneath another.

submersible: a small research submarine, requiring a support ship.

syncline: a downward fold of rock layers.

tephra: pyroclastic debris.

transform fault: a type of strike-slip fault, usually crossing the Mid-Ocean Ridge at a right angle and offsetting one segment of the Ridge from the next.

translation: horizontal movement of one block of crust relative to another along a fault.

travertine: a calcium carbonate mineral deposited where groundwater is exposed to air: usually at hot springs, and in caves as stalactites and stalagmites.

trona: a mineral often formed by evaporation and found in saline lake residue; a major source of sodium compounds (soda).

tsunami: a seismic sea wave, caused by an earthquake, undersea earthslide, subsidence, or volcanic eruption; often inaccurately called a tidal wave.

tuff: tephra solidified into layers of rock.

unconformity: discontinuity in rock layers, where one type of rock changes abruptly to another, showing a gap in the record.

Uniformitarianism: the doctrine that geologic processes have acted in the same regular manner throughout geologic time.

Eon	Era	Period	Epoch	MYA
Phanerozoic	Cenozoic	Quaternary	Holocene	0.010
			Pleistocene	2
		Tertiary	Pliocene	5
			Miocene	24
			Oligocene	38
			Eocene	55
			Paleocene	63
	Mesozoic	Cretaceous		138
		Jurassic		205
		Triassic		240
	Paleozoic	Permian		290
		Carboniferous	Pennsylvanian	330
			Mississippian	360
		Devonian		410
		Silurian		435
		Ordovician		500
		Cambrian		570
Proterozoic	Precambrian time			2,500
Archean				3,800
				4,600

This chart reflects U. S. Geological Survey usage. Figures in the MYA (millions of years ago) column approximate the beginning of each time division.

Illustrations
Credits

358   Abbreviations: (t)—top; (c)—center; (b)—bottom; (r)—right; (l)—left NGS—National Geographic Staff; NGP—National Geographic Photographer; NGA—National Geographic Publications Art; NGC—National Geographic Cartography; SRW—Stansbury, Ronsaville, Wood Inc.

Cover stamping by Dynamic Graphics, Inc. and David Seager, NGS. Pages 2-3, Ralph Perry. 4-5, S. Jónasson, Bruce Coleman, Inc. 6-7, Devon Jacklin. 8-9, Annie Griffiths. 14-15, Maurice and Katia Krafft.

Discovery of Planet Earth
16-17, Stan Osolinski. 19, John Clerk, from "James Hutton's Theory of the Earth: The Lost Drawings," Scottish Academic Press, 1978. 20-25, Rob Wood, SRW. 26-27, Phil Jordan, Beveridge & Associates, Inc. 28, NGA. 29, Steve Raymer, NGP. 30, Wayne McLoughlin. 31-35, Kenneth Garrett. 37, Paul M. Breeden and NGA. 38, Rob Wood, SRW. 40-41, NGA. 42, Wayne McLoughlin. 43, Denis Serrette, Institut de Paleontologie M.N.H.N. Paris. 44-45, Rick Smolan, Contact Press Images. 46-47, 48-49, 50-51, 52-53, Robert Hynes. 47(r), 49(r), 51(r), 53(r), Paul M. Breeden. 55, Gordon W. Gahan. 56, David Muench. 58-59, NGA.

Spreading
60-61, Emory Kristof, NGP. 62, Rob Wood, SRW. 63(t), NGA. (b), Rob Wood, SRW.

Birth of An African Ocean
64-65, Emory Kristof, NGP. 66, NGC. 67, George F. Mobley, NGP. 68-69, Jay H. Matternes. 69, Bob Campbell. 70, Robert Harding Picture Library. 71-79, George F. Mobley, NGP. 80-81, Emory Kristof, NGP. 81(t), Alan Root. 82-83, Dynamic Graphics, Inc. and Rob Wood, SRW. 84-85, Art Wolfe. 85(t), George F. Mobley, NGP. 86-87, Emory Kristof, NGP. 88, Volkmar Wentzel, NGP. 89, NGA. 90-91, Emory Kristof, NGP. 92-93, Georg Gerster. 94-95, Marion Kaplan. 96, Emory Kristof, NGP. 97, NGA. 98-99, Maurice & Katia Krafft. 100-101, Eliot Porter.

The Mountains of the Sea
102-103, Alvin M. Chandler and Emory Kristof, both NGS. 104-105, O. Louis Mazzatenta, NGS. 106, NGC. 107-109, John A. Bonner, NGC; data courtesy of GEBCO charts from Canadian Hydrographic Service, Ottawa; International Hydrographic Organization, Monaco; and UNESCO. 111, Emory Kristof, NGP. 112-115, Davis Meltzer. 116,

Woods Hole Oceanographic Institution (WHOI). 117, Ken C. McDonald. 118-119, Davis Meltzer. 120, Kientzy and Suteau, CNEXO. 121, Leroux and Guerrero, CNEXO. 122, Robert Hynes. 124-125, Fred Grassle, WHOI, courtesy of Robert Hessler, Scripps Institution of Oceanography. 126-127, WHOI. 128-129, Ruth Turner, courtesy Robert Hessler, Scripps. 130, R. Hekinian, CNEXO, Centre Oceanologique de Bretagne. 131-133, Emory Kristof, NGP.

Land of Ice and Fire
134-135, Randall Hyman. 136(t), NGC. (b), NGA. 137, Randall Hyman. 138-139, Maurice & Katia Krafft. 140-141, Chris Foss. 142-143, Sigurdur Thórarinsson. 144-148, Randall Hyman. 150-155, Maurice & Katia Krafft. 156-157, Robert S. Patton, NGS.

Hotspots
158-159, Maurice & Katia Krafft. 160, Rob Wood, SRW. 161(t), NGA. (b), Rob Wood, SRW.

Earth's Fountains of Heat
162-163, Steve Raymer, NGP. 165(t), Dick Durrance II. (b), Maurice & Katia Krafft. 166(t), NGC. (b), Jaime Quintero and Paul M. Breeden. 167, Jonathan Blair. 168-169, Robert W. Madden, NGP. 170-171, Norman G. Banks, USGS. 172-173, Paul Chesley. 174-175, Maurice & Katia Krafft. 176, David Muench. 177-179, Gordon W. Gahan. 180-181, Tui De Roy Moore. 182-183, Sally Anne Thompson, Animal Photography. 184-185, Fiip Schulke, Black Star. 187, James Balog, Black Star. 188, NGA. 189, Doug Peacock. 190(b), Wayne McLoughlin. 190-191, Dynamic Graphics, Inc. and Rob Wood, SRW. 192-193, Steven Fuller. 194-195, Dick Durrance II. 196-201, Steven Fuller. 202-203, Entheos.

Slipping
204-205, Georg Gerster. 206, Rob Wood, SRW. 207(t), NGA. (b), Rob Wood, SRW.

On the Grindstone Edges
208-209, Robert W. Madden, NGP. 210, W. E. Garrett, NGS. 212(t), NGC. (b), NGA. 213-215, David Doubilet. 216-217, Nathan Benn. 218-219, Kazuyoshi Nomachi, Tuttle-Mori Agency, Inc. 220-221, Nathan Benn. 222(t), NGA. 222-223, Emory Kristof, NGP. 224-225, Gordon W. Gahan. 226, James A. Sugar. 227, NGA. 228, Wayne McLoughlin. 229, James A. Sugar. 230-231, The Bancroft Library, University of California. 232-233, The LeBaron Collection. 234-235, California Historical

Society, San Francisco. 236, Philip R. Leonhardi, NGS and James A. Sugar. 237, James P. Blair, NGP. 238-239, Dynamic Graphics, Inc. and Rob Wood, SRW. 240-243, James A. Sugar. 245(t), Davis Meltzer. (b), Philip R. Leonhardi, NGS and James A. Sugar. 246-247, James A. Sugar.

Collision
248-249, Thomas J. Abercrombie, NGS. 250, Rob Wood, SRW. 251(t), NGA. (b), Rob Wood, SRW.

The Ring of Fire
252-253, Steve Raymer, NGP. 255, NGA. 256-259, Chris Foss. 260-261, William Albert Allard. 262-263, Dynamic Graphics, Inc. and Rob Wood, SRW. 264-265, George F. Mobley, NGP. 266-267, Ralph Perry. 268-269, Douglas Miller, West Stock, Inc. 270-271, Robert W. Madden, NGP. 272, John Marshall. 273, Ron Cronin. 274-275, Robert W. Madden, NGP. 275(t), John Marshall. 276-277, Steve McCutcheon, Alaska Photo. 278, Stern, Black Star. 280-281, James L. Stanfield, NGP. 282-283, Michael S. Yamashita. 284-285, Shin Yoshino, Orion Press. 286, Maurice & Katia Krafft. 287, Michael S. Yamashita. 288-289, Gordon W. Gahan.

Ashes and Empires
290-291, David Austen. 292(t), NGA. (b), NGC. 293, Maurice & Katia Krafft. 294-297, David Austen. 298-299, Maurice & Katia Krafft. 300-303, David Austen. 304-305, Guillermo Aldana E. 305(t), NGA. 306 (t), Wayne McLoughlin. (b), Bruce E. Hunter. 307(t), NGA. (b), Otis Imboden, NGP. 308-314, Nathan Benn. 316-317, Gordon W. Gahan. 317, Otis Imboden, NGP.

When Continents Collide
318-319, Jim Brandenburg. 320, David Muench. 321(t), Rob Wood, SRW. (b), NGA. 322-323, T. W. Hall. 324-325, Paul Chesley. 326(t), NGC. (b), Nathan Benn. 328, Jean-Paul Ferrero, Ardea London. 329, Nathan Benn. 330-331, Thomas J. Abercrombie, NGS. 332-333, Nathan Benn. 334-335, Yoshikazu Shirakawa, Image Bank. 336, Emory Kristof, NGP. 337, Paläontologisches Institut, Zurich, Nathan Benn. 338, Christian Bonington, Daily Telegraph Colour Library. 338-339, Jonathan Blair. 340-341, Gianni Tortoli. 342, G. Rancinan, Sygma. 344-345, Davis Meltzer. 346-347, Yoshikazu Shirakawa, Image Bank. 348-349, Reinhold Messner. 350-351, Galen Rowell. 352-353, Harold A. Knutson. 354-355, Scott Rowed.

Bibliography and Acknowledgments

We found the following books and articles particularly helpful. Many are intended for nonspecialist audiences.

Baker, B. H. *Geology of the Eastern Rift System of Africa*. Boulder: Geological Society of America, 1972.

Bates, Robert L., and Julia A. Jackson, eds. *Glossary of Geology*. Falls Church, VA: American Geological Institute, 1980.

Beadle, L. C. *The Inland Waters of Tropical Africa*. London: Longman, 1981.

Bolt, Bruce A. *Earthquakes: A Primer*. San Francisco: W. H. Freeman & Co., 1978.

———. et al. *Geological Hazards*. New York: Springer Verlag, 1975.

Bullard, Fred M. *Volcanoes of the Earth*. Austin: U. of Texas Press, 1976.

Burke, Kevin C., and J. T. Wilson. "Hot Spots on the Earth's Surface." *Scientific American* 235 (August 1976): 46-57.

Continents Adrift and Continents Aground: Readings from Scientific American. San Francisco: W. H. Freeman & Co., 1976.

Dalrymple, G. Brent et al. "Origin of the Hawaiian Islands." *American Scientist* 61 (6 May 1973): 294-307.

Decker, R., and B. Decker. *Volcanoes*. San Francisco: W. H. Freeman & Co., 1981.

Dietrich, Günter et al. *General Oceanography*. New York: John Wiley & Sons, 1980.

Dott, R. H., and R. L. Batten. *Evolution of the Earth*. New York: McGraw-Hill, 1981.

Eicher, Don L., and A. Lee McAlester. *History of the Earth*. Englewood Cliffs, NJ: Prentice-Hall, 1980.

Fairbridge, Rhodes W. *Encyclopedia of Oceanography*. New York: Reinhold, 1966.

Frakes, Lawrence A. *Climates Throughout Geologic Time*. New York: Elsevier, 1979.

Francheteau, Jean et al. *CYAMEX: Naissance d'un Océan*. Centre National Pour L'Exploitation des Océans, 1980.

Hadley, David M. et al. "Yellowstone: Seismic Evidence for a Chemical Mantle Plume." *Science* 193 (24 September 1976):1237-1239.

Hallam, A. *A Revolution in the Earth Sciences*. Oxford: Clarendon Press, 1973.

———, ed. *Planet Earth: An Encyclopedia of Geology*. Oxford: Elsevier-Phaidon, 1977.

Holmes, Arthur. *Holmes Principles of Physical Geology*. New York: John Wiley & Sons, 1978.

Horsfield, Brenda, and Peter Bennet Stone. *The Great Ocean Business*. New York: Coward, McCann & Geoghegan, 1972.

Iacopi, Robert. *Earthquakes*. Menlo Park, CA: Lane Publishing Co., 1971.

James, David L. et al. "The Growth of Western North America." *Scientific American* 247 (November 1982): 70-84.

Keefer, William R. *The Geologic Story of Yellowstone National Park*. Washington, D. C.: Government Printing Office, 1971.

Keller, Werner. *The Bible As History*. New York: William Morrow, 1956.

Levin, Harold L. *The Earth Through Time*. Philadelphia: W. B. Saunders Co., 1978.

MacDonald, Gordon A. *Volcanoes*. Englewood Cliffs, NJ: Prentice-Hall, 1972.

Marvin, Ursula B. *Continental Drift*. Washington, D. C.: Smithsonian Press, 1973.

McPhee, John. *Basin and Range*. New York: Farrar, Straus, Giroux, 1981.

Molnar, Peter, and Paul Tapponnier. "The Collision between India and Eurasia." *Scientific American* 236 (April 1977): 30-41.

Moore, James G. "Mechanism of Formation of Pillow Lava." *American Scientist* 63 (May/June 1975): 269-277.

Motz, Lloyd. *Rediscovery of the Earth*. New York: Van Nostrand Reinhold Co., 1975.

Press, Frank, and Raymond Siever. *Earth*. San Francisco: W. H. Freeman & Co., 1982.

Ritchie, David. *Ring of Fire*. New York: Atheneum, 1981.

Ross, David A. *Introduction to Oceanography*. Englewood Cliffs, NJ: Prentice-Hall, 1982.

Seyfert, Carl K., and Leslie A. Sirkin. *Earth History and Plate Tectonics*. New York: Harper & Row, 1979.

Sheets, Payson D., and Donald K. Grayson, eds. *Volcanic Activity and Human Ecology*. New York: Academic Press, 1979.

Simkin, Tom et al. *Volcanoes of the World*. Stroudsburg, PA: Hutchinson Ross, 1981.

Smith, David G., ed. *Cambridge Encyclopedia of Earth Sciences*. New York: Cambridge U. Press, 1982.

Smith, R. B., and Robert L. Christiansen. "Yellowstone Park as a Window on the Earth's Interior." *Scientific American* 242 (February 1980): 104-117.

Sullivan, Walter. *Continents in Motion*. New York: McGraw-Hill, 1974.

Vitaliano, Dorothy. *Legends of the Earth*. Bloomington: Indiana U. Press, 1973.

Williams, Howel, and Alexander R. McBirney. *Volcanology*. San Francisco: Freeman, Cooper & Co., 1979.

Windley, Brian F. *The Evolving Continents*. New York: John Wiley & Sons, 1977.

Wyllie, Peter J. *The Way the Earth Works*. New York: John Wiley & Sons, 1976.

Young, Patrick. "The Earth's Fountains of Fire." *Mosaic* 11 (May/June 1980): 2-9.

Ziegler, A. M., and C. R. Scotese, eds., *Paleogeographic Atlas*. Chicago: U. of Chicago Press. Vol. 1 scheduled for publication in 1984 (to incorporate several previously published articles).

We made extensive use of the following periodicals: *Earthquake Information Bulletin; EOS; Geology; Geotimes; Journal of Geophysical Research; National Geographic Magazine; Natural History; Nature; New Scientist; Oceanus; Science; Science 80, 81, 82; Science News; Scientific American*.

We wish to thank the many scientists and consultants who generously contributed their time and knowledge.

U. S. Geological Survey. Reston, Va.: Harold Burstyn, George Ericksen, John Filson, Walt Hays, Glen Izett, Michael Ryan, Richard Williams, Wallace de Witt. Menlo Park, Calif.: Robert Brown, Robert Christiansen, Darrell Herd, James G. Moore, George Plafker, Sandra Schultz. Elsewhere: Rick Garrison, Robin Holcomb, Kim Klitgord, Don Peterson, Richard Powers, Dorothy Vitaliano.

Universities. Brian Baker, U. of Oregon; John Baross, Oregon State; Jelle de Boer, Wesleyan U.; Clark Burchfiel, MIT; Kevin Burke, State U. of New York at Albany; Harmon Craig, UC San Diego; Thomas Crough, Purdue; Clifford Dahm, Oregon State; Charles Drake, Dartmouth; Robert Duncan, Oregon State; John Edmond, MIT; Clifford Hopson, UC Santa Barbara; Gary Johnson, Dartmouth; Bruce Marsh, Johns Hopkins; Lisa McBroome, U. of Hawaii; Hugh McNiven, UC Berkeley; Peter Molnar, MIT; Jason Morgan, Princeton; Amos Nur, Stanford; Ray Pestrong, San Francisco State; William Schopf, UCLA; John Sclater, MIT; Christopher Scotese, U. of Chicago; Payson Sheets, U. of Colorado; Haraldur Sigurdsson, U. of Rhode Island; Peter Stifel, U. of Maryland; Half Zantop, Dartmouth.

Woods Hole Oceanographic Institution. W. B. Bryan, John Donnelly, Frederick Grassle, J. R. Heirtzler, Holger Jannasch.

Smithsonian Institution. Nicholas Hotton, Francis Hueber, Lindsey McClellan, Robert Purdy, Tom Simkin. *Other institutions; U. S. government; consultants abroad*. Zvi Ben-Avraham, Jean-Louis Cheminée, Richard De Rycke, John Dewey, Páll Einarsson, Jean Francheteau, Gianni Frazzetta, Ben Gadd, Roderick Hutchinson, Ken Iten, David James, Leonard Johnson, Igor Loupekine, Paul Lowman, Stephen Maran, David Monahan, James Monger, Antonio Rapolla, James Sedell, Konrad Staudacher, Paul Tapponnier, Sigurdur Thórarinsson, Lillo Villari, George Wetherill.

359

Index

Illustrations appear in **boldface** type, caption references in *italic*, and text in lightface.

Type composition by National Geographic's Photographic Services. Color separations by Beck Engraving Co., Inc., Philadelphia, Pa.; Chanticleer Co., Inc., New York, N.Y.; The Lanman Progressive Companies, Washington, D. C.; Offset Separations Corp., New York, N.Y.; Scan Studios Limited, Dublin, Ireland. Printed and bound by R. R. Donnelley & Sons Co., Chicago, Ill. Paper by Mead Paper Co., New York, N.Y.

Library of Congress CIP Data

Ballard, Robert D.
   Exploring our living planet.

   Bibliography: p.
   Includes index.
   1. Plate tectonics.   I. National Geographic Book Service.   II. Title.
QE511.4.B34   1983   551.1'36   83-2336
ISBN 0-87044-459-X
ISBN 0-87044-397-6 (lib. bdg.)
ISBN 0-87044-460-3 (deluxe)